Highway Capacity and Quality of Service 2018

Transportation Research Record: Journal of the Transportation Research Board

Volume 2672, Issue 15

ISBN: 978-1-5108-7733-7

Printed from e-media with permission by:

Curran Associates, Inc.
57 Morehouse Lane
Red Hook, NY 12571

Some format issues inherent in the e-media version may also appear in this print version.

Published by Sage Publications USA

ISBN (Print) 978-1-5108-7733-7
ISBN 2018 Printed Set (All Issues) 978-1-5108-7735-1

Printed by Curran Associates, Inc. (2018)

Additional copies of this publication are available from:

Curran Associates, Inc.
57 Morehouse Lane
Red Hook, NY 12571 USA
Phone: 845-758-0400
Fax: 845-758-2633
Email: curran@proceedings.com
Web: www.proceedings.com

Volume 2672 Number 15 December 2018

TRR
JOURNAL OF THE TRANSPORTATION RESEARCH BOARD

Contents

Highway Capacity and Quality of Service 2018

Articles

Article

TRR
JOURNAL OF THE TRANSPORTATION RESEARCH BOARD

Impact Assessment of Short-Term Work Zones on Intersection Capacity in New York City

Transportation Research Record
2018, Vol. 2672(15) 1–8

DOI: 10.1177/0361198118774703
journals.sagepub.com/home/trr
SAGE

Nawaf M. Alshabibi[1] and Elena Prassas[2]

Abstract

New York City infrastructure is one of the oldest transportation infrastructures in the United States. Local street construction and short-term work zones are almost continuously planned events that affect the movement of traffic on city streets by requiring the closing of one or more lanes at intersections throughout NYC, and it is important to understand the effect on capacity due to such work. This paper looks at the effect of short-term work zones on the capacity of signalized intersections in New York City. Data was collected at five locations in New York City, both during the work zone and then again after the work zone was removed. Over 25 hours of video data was collected and reduced. It was found that at all locations, the saturation headway was smaller during the work zone compared to after the work zone was removed, that is, the saturation flow rate per lane increased during the work zone. This was an unexpected result. A possible reason for this is the increased traffic pressure that drivers feel when a lane is closed. Thus, although overall approach capacity does decrease because a lane is closed, it did not decrease as much as expected. The field values are then compared with those from two other models: the Highway Capacity Manual model and a model developed by Schroeder et al. It was found that both models underestimate the capacity of the approach.

The estimation of capacity at signalized intersections is an ongoing process for departments of transportation (DOTs) around the country. It is important for predicting queues and delay and for making decisions on signal timing. Delay and congestion impose large costs to both the construction contractors and the motor vehicles. The effect of lane closures due to work zones has a significant effect on capacity and thus the mobility of road users.

New York City (NYC) infrastructure is one of the oldest transportation infrastructures in the United States. Local street construction and short-term work zones are almost continuously planned events that affect the movement of traffic on city streets by requiring the closing of one or more lanes at intersections throughout NYC. When a work zone is needed at a location, temporary traffic control plans (TTCPs) are developed to promote the most efficient movement of traffic. Knowing the capacity of the approach is vital to this task.

This paper is a case study of the effect of work zones on saturation headway, and thus capacity, in NYC. The paper specifically looks at the effect of short-term work zones on the capacity of signalized intersections in NYC. Data was collected at five locations in NYC, both during the work zone and then again after the work zone was removed. Over 25 hours of video data was collected and reduced. The average saturation flow rate found from the field data are then compared with saturation flow rate calculated from two other models, the Highway Capacity Manual (HCM) model and the model in Reference (*1*). The HCM (*2*) model is the most widely used model and the second model tested is a more recent model developed by Schroeder et al. (*1*). The purpose of this study was to test the existing models to see if any can reliably be used in NYC.

Literature Review

Much research has been done concerning the effect of work zones on freeway saturation flow rate (*3*–*7*). However, this is not the case for the effect of work zones on intersection saturation flow rate. In fact, only five such references could be found.

Recent research on the effect of work zones on intersection saturation flow rate was done by Bonneson and Nguyen (*8*) and was incorporated into the latest version of the HCM published in 2016 (*2*). The literature review in this reference (*8*) found only three previous references on the effect of

[1]Department of Urban and Regional Planning, Imam Abdulrahman Bin Faisal University, Dammam, Saudi Arabia
[2]Department of Civil and Urban Engineering, NYU Tandon School of Engineering, Brooklyn, NY

Corresponding Author:
Address correspondence to Nawaf M. Alshabibi: nma295@nyu.edu

work zones on capacity in urban streets (*9–11*). After an extensive literature search, only one additional reference was found concerning work zones on urban streets (*1*). The following is a description of each of these five references.

Hawkins (*9*) found capacity reduction due to the presence of a work zone when examining one arterial in Texas. The work zone was midblock and saturation flow rate was measured through the work zone (not at the intersection). Capacity was found to be 760 vphpl (vehicles per hour per lane). There was no measurement of saturation flow rate when no work zone was present, however, to isolate the effect of the work zone.

In 1988, Joseph et al. (*10*) developed a computer program to evaluate the effect of work zones on arterials. The model used a combination simulation and macroscopic model to predict the delay at the intersection. Saturation flow rate, however, was an input to this model. The factors found most influential were platoon cohesion, signal timing and the location of the work zone on the arterial.

Using simulation, Elefteriadou et al. (*11*) created regression equations to predict capacity at an intersection when a work zone is present. It was found that geometry, distance from intersection, and green to cycle length (*g/C*) ratio were the major factors impacting capacity. It was also found that under congested conditions, capacity may increase, particularly when there are more than 3 lanes on the approach. The simulation showed turning vehicles getting blocked behind the construction, allowing other vehicles to flow more freely. It is unclear at what level of congestion this occurred since no value of volume to capacity (v/c) ratio was reported.

The Bonneson and Nguyen work (*8*) measured intersection headways both during the work zone and after the work zone was removed. Data was collected at seven locations. All the sites for which data were collected were on two-way intersections in non-central business district (CBD) areas and all had exclusive left-turn lanes. Five of the seven sites for which they had both during and after data had measured saturation headways that were larger during the work zone. Two of the sites had lower headways during the work zone compared to after the work zone. The headways were then adjusted to remove the effect of heavy vehicles and right-turning vehicles. On average, the saturation headway during the work zone was 5.2% higher than without the work zone.

From this data (*8*), a model was developed to calculate an additional factor to the saturation flow rate calculation in the HCM, that is, as a multiplicative factor for adjusting the base saturation flow rate for the presence of the work zone at the intersection. The factor reduces saturation flow rate based on the total width of all the open lanes during the work zone, the number of lanes during the work zone, and the number of lanes when there is no work zone. The maximum work zone factor is limited to 1.0.

Equations (1), (2), and (3) show the equations calibrated to calculate the adjustment factor to saturation flow rate at an intersection due to the presence of a work zone.

$$f_{\text{wz}} = 0.858 f_{\text{wid}} f_{\text{reduce}} \leq 1.0 \quad (1)$$

$$f_{\text{wid}} = \frac{1.0}{1.0 - 0.0057(a_{\text{w}} - 12)} \quad (2)$$

$$f_{\text{reduce}} = \frac{1.0}{1.0 - 0.0402(n_o - n_{\text{wz}})} \quad (3)$$

where

f_{wz} = adjustment factor for work zone presence,
f_{wid} = adjustment factor for approach width,
f_{reduce} = adjustment factor for lanes reduced during work zone,
a_w = total approach width during work zone,
n_o = number of lanes open under normal conditions, that is, no work zone present, excluding right-turn only lanes, and
n_{wz} = number of lanes open when work zone is present, excluding right-turn only lanes.

Reference (*1*), by Schroeder et al., is the one recent paper found from an extensive literature search that examined the issue of work zones at signalized intersections after the HCM model was created. Schroeder et al. collected data at five intersections in North Carolina. As in the previous research, all the sites were on two-way non-CBD arterials. All the locations had either exclusive turning lanes or very low volumes of turning vehicles. Multiple linear regression was used to create two additive models for predicting headway during a work zone. An additive model adjusts the base headway by adding and subtracting values due to the influence of relevant variables. In these models, the relevant variables were percentage of right and left turns, closed or open exclusive left or right turn lanes, work intensity, and pavement condition.

The multiple linear regression model created and recommended to predict saturation headway during a work zone is shown in Equation (4).

$$h_s^{\text{adj}} = 2.04 + 0.136(LWI) + 0.271(HWI) + 0.211(PC) + 0.112(PL) + 0.003(TTP) + 0.055(EX) \quad (4)$$

where

LWI = work intensity low (0: no; 1: yes),
HWI = work intensity high (0: no; 1: yes),
PC = pavement condition (0: normal; 1: milled),
PL = presence of ledge (0: no; 1: yes),
TP = turn percentage (combined left and right-turn percent), and
EX = number of closed exclusive lanes (combined left and right).

The left and right turn percentage values were combined because when included separately, they did not have a significant impact. The same was true of exclusive left and right turn lanes.

The 2.04 in Equation 4 represents the base saturation headway. This was found as the average thru lane headway of the

Table 1. Work Zone Locations and Characteristics

Site	Type intersection work zone approach/cross street	Lanes during work zone	Lanes closed	Hours of data	Lanes after work zone	Hours of data
1	One-way work zone approach northbound/two-way cross street	4	Right curb turn lane	5	5	5
2	Two-way work zone approach eastbound/one-way southbound	2	Left-most thru lane	5	3	5
3	One-way work zone approach southbound/one-way eastbound	3	Right curb thru lane	1.5	4	1.5
4	One-way work zone approach northbound/two-way	2	Left curb lane	3	3	2
5	One-way work zone approach westbound/one-way southbound	1	Thru lane	2	2	2

five sites. This number can be substituted with a locally calibrated base saturation flow rate or a default value, such as 1.89 from the HCM. (The HCM uses an ideal saturation flow rate of 1,900 passenger cars per hour green per lane [pcphgpl], which is equivalent to a headway of 1.89 seconds per vehicle [spv].)

Data Collection and Reduction

The data for this study was collected at six intersections in Manhattan. Only five sites had usable data because queues were not long enough to measure average saturation headway at the sixth intersection. Table 1 summarizes the site locations and characteristics. The video data was recorded by the NYCDOT Traffic Management Center (TMC) of the approach to the intersection that contained the work zone, both during the work zone and after the work zone was removed.

All the approaches had the following similar characteristics.

- CBD location
- No exclusive turn bays or lanes
- 90-second cycle length
- Short-term work zone closure
- Work zone within 250 feet of the intersection
- Headways measured during unsaturated conditions, but with significant queues of at least 6 to 8 vehicles per cycle
- Lane widths between 11 and 13 feet, thus not affecting saturation flow rate, as defined in the HCM (*2*).
- No grade
- No parking at the time of data collection
- No buses stopping within 250 feet of the intersection
- No bicycles

The type of data reduced from the videos, by cycle, included:

- Prevailing saturation headway during and after the work zone before the first heavy vehicle arrived (that is, only queues without heavy vehicles were used to remove the effects of heavy vehicles from saturation headway)
- Start-up lost time
- Percent turning vehicles in shared lanes

Saturation headway per cycle was found from the average headway of vehicles in the queue after the fourth queued vehicle. Most of the cycles used had 8 vehicles in queue, although a few cycles had 6 or 7 vehicles. Saturation flow rate was calculated using the method described in References (*2*), (*8*) and (*12*), that is, by finding the saturation headway per cycle and averaging those to find the average saturation headway. Saturation flow rate is then found by dividing the 3,600 seconds per hour by the average saturation headway.

Data Summary

As stated earlier, only queues without heavy vehicles were used in this study.

Table 2 summarizes the data for the thru-only lanes during and after the work zone. The thru-only lane data after the work zone was removed represents the saturation headway under ideal conditions. The only difference between the "during work zone" conditions and the "after work zone" conditions at each site is the presence of the work zone, thus isolating the effect of the work zone. All the thru-only lanes have saturation headways that are lower during the presence of the work zone. Lower saturation headways translate to higher saturation flow rates and thus higher capacities in these lanes. This is the opposite result from what was expected, and opposite what is predicted in the HCM (*2*). It is possible that in high-density, high-pedestrian volume CBD areas like NYC, there is more intense traffic pressure on vehicles during the work zone to keep headways smaller.

Table 2. Thru-Only Lane Data, During and After Work Zone

Site	Number samples during work zone	Average prevailing saturation headway	Standard deviation	Number samples after work zone	Base saturation headway	Standard deviation	t-stat	*P*-value
1	173	3.00	0.687	258	3.48	0.876	−6.132	9.8E-10
2	98	2.52	0.607	208	3.08	1.079	−4.837	1.05E-06
3	19	2.27	0.469	78	2.66	0.605	−2.63	0.005
4	n/a	n/a	n/a	21	2.94	0.484	n/a	n/a
5	n/a	n/a	n/a	34	2.39	0.506	n/a	n/a

Note: n/a = not available

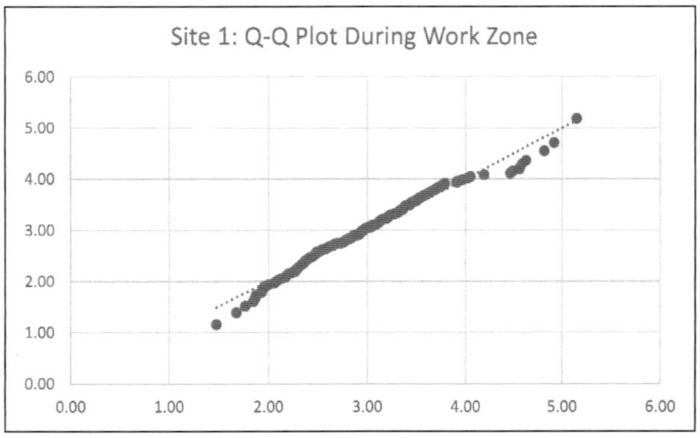

Figure 1. Q-Q Plot of Site 1 headway data during the work zone.

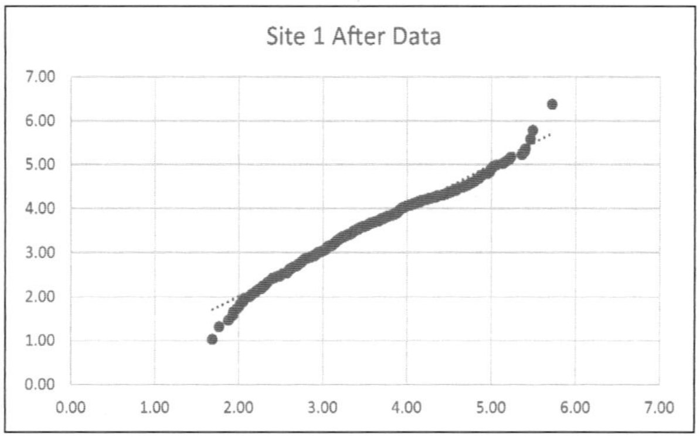

Figure 2. Q-Q Plot of Site 1 headway data after the work zone.

The data was statistically-analyzed and results showed that saturation headways are approximately normally distributed, as also found in References (*13–17*). For example, Figures 1 and 2 show the Q-Q plots of the data against a standard normal distribution for Site 1 during and after the work zone, respectively. The points are roughly a straight line and can be assumed to be from the same distribution. Figure 3 clearly shows the shift in the mean between the during- and after-cycle average headway data.

At sites 4 and 5, there is only one thru-only lane which then becomes a shared lane during the work zone, and thus a comparison cannot be made. For the remaining sites, the thru-only lanes showed a significant decrease in saturation headway per lane during the work zone.

To compare the headways in the shared lanes at each site, the effect of turning vehicles and pedestrians had to be removed. The HCM turn factors and pedestrian-interference factor were used. The headways in the shared lanes after removing these effects are very low, as can be seen in Tables 3 and 4. Examining the HCM factors, it was found that the pedestrian adjustment factors were extremely low. This may be due to conditions under which this factor was calibrated and may not be accurate under NYC conditions. For the comparison of during and after headways, however, since the same factors are used to adjust all headways, the relative difference still represents the effect of the work zone.

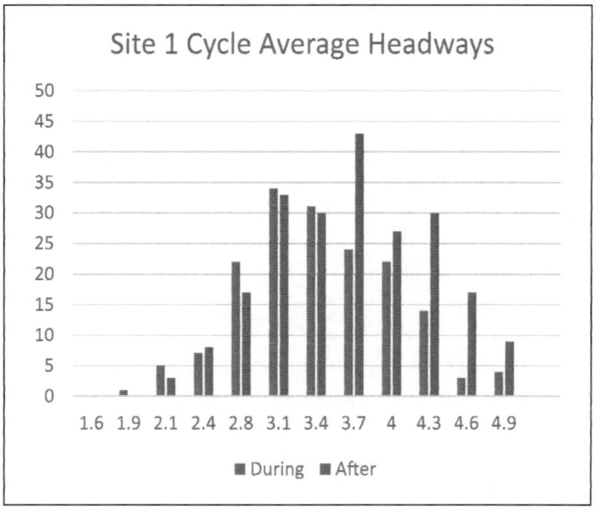

Figure 3. Headways at Site 1.

The shared lane data in Table 3 and Table 4 are separated by the location of the shared lane relative to the work zone. The data showed different results where the work zone was

Table 3. Shared-Lanes Not Adjacent to Work Zone

Site #: Lane type	Number samples during work zone	Adjusted average prevailing saturation headway during work zone	Standard deviation	Number samples after work zone	Adjusted average saturation headway after work zone	Standard deviation	*t*-value	*P*-value
1: LT/TH	51	2.00	0.506	20	2.26	0.533	−1.88	0.03
2: TH/RT	46	2.02	0.802	7	2.09	0.682	−0.01	0.41
3: LT/TH	7	1.94	0.217	7	2.39	0.587	−1.045	0.15
4: TH/RT	19	1.46	0.505	4	1.98	.2590	−1.948	0.03

Note: LT = left-turn; TH = thru-lane; RT = right-turn.

Table 4. Shared orTurning-Lane Adjacent to Work Zone

Site#: Lane type	Number samples during work zone	Adjusted average prevailing saturation headway during work zone	Standard deviation	Number samples after work zone	Adjusted average saturation headway after work zone	Standard deviation	*t*-value	*P*-value
1: TH/RT	65	1.65	0.383	88	1.69	0.437	0.674	0.25
4: LT/TH	43	1.30	0.316	32	1.38	0.33	1.072	0.14
5: LT/TH	34	2.02	0.689	n/a	n/a	n/a	n/a	n/a
5: Default LT only	n/a	n/a	n/a	28	3.07	0.91	n/a	n/a

Table 5. HCM Work Zone Factors

Site	Approach width during work zone	Open lanes without work zone	Open lanes with work zone	Hcm work zone factor, f_{wz}
1	45	5	4	1.000
2	24	3	2	0.885
3	35	4	3	0.949
4	23	3	2	0.880
5	11	2	1	0.820

adjacent to the shared lane compared to sites where the work zone was not adjacent to the shared lane.

Table 3 summarizes the data for the shared lanes that are not adjacent to the work zone. There was a significant decrease seen in the saturation headway during the work zone at Sites 1 and 4. There was no significant decrease at Sites 2 and 3. Sample sizes in these lanes were low and thus the results are inconclusive.

Table 4 summarizes the data for the shared lanes adjacent to the work zone. It can be seen that there was no significant difference found in the saturation headway during the work zone in the lane that was next to the work zone. The sample sizes at these two locations were relatively large and it is possible that these two sites suggest that the lane next to the work zone does not experience a change in saturation flow rate.

Change in saturation headway in the shared lanes was mixed. The two shared lanes that were adjacent to the work zone showed no significant difference in saturation headway. Two of the four shared lanes that were not adjacent to the work zone had a significant decrease in saturation headway. The remaining two had very low sample sizes.

From this measured data, it appears that the per lane saturation headway either decreases or remains the same during a work zone depending on the type of lane and its location compared to the work zone. The capacity of the approach, of course, does go down due to losing a lane, but not as much as expected.

Comparison with Other Models

Because the HCM is the most commonly used methodology for estimating signalized intersection capacity, the measured values in the data are compared to the HCM values calculated using Equations (1)–(3). Table 5 shows the work zone factor calculated using the HCM methodology.

The HCM estimated saturation flow rate is calculated using the measured "base" saturation headway after the work

Table 6. Measured and HCM Lane Group Saturation Flow Rate per Lane

Site	Lane group	Measured "Base" headway after work zone spv	Work zone factor f_{wz}	Hcm (2) saturation flow during vphgpl	Schroeder et al. model saturation flow (1)	Adjusted measured headway during work zone s/veh	Data saturation flow during vphgpl
1	TH	3.48	1.000	**1,034**	**965**	3.00	**1,200**
1	LT/TH	2.26	1.000	**1,593**	**1,435**	2.00	**1,800**
1	TH/RT	1.69	1.000	**2,130**	**1,858**	1.69	**2,130**
2	TH	3.08	0.885	**1,035**	**1,082**	2.52	**1,429**
2	TH/RT	2.09	0.885	**1,525**	**1,540**	2.09	**1,722**
3	TH	2.66	0.949	**1,285**	**1,238**	2.27	**1,586**
3	LT/TH	1.37	0.949	**2,494**	**2,225**	1.37	**2,628**
4	LT/TH	1.38	0.880	**2,296**	**2,211**	1.38	**2,609**
4	TH/RT	1.98	0.880	**1,600**	**1,616**	1.46	**2,466**
5	LT/TH	n/a	0.820	**n/a**	**n/a**	2.02	**1,782**

Note: Bold values are saturation flow rates.

Table 7. Comparison of Measured, HCM, and Schroeder et al. Lane Group Capacities during the Work Zone

Site	Lane group	N	g/C	Data lane group capacity during	HCM lane group capacity during	Schroeder et al. [11] lane group capacity	Data approach capacity	HCM approach capacity	Schroeder approach capacity
1	TH	2	0.5	1200	1034	932	3165	2895	2612
1	LT/TH	1	0.5	900	796	681			
1	TH/RT	1	0.5	1065	1065	868			
2	TH	1	0.33	476	345	343	1050	853	865
2	TH/RT	1	0.33	574	508	480			
3	TH	2	0.28	881	713	662	1610	1405	1316
3	LT/TH	1	0.28	729	692	575			
4	LT/TH	1	0.5	1304	1147	1021	2542	1947	1913
4	TH/RT	1	0.5	1233	800	761			
5	LT/TH	1	0.5	891	n/a	n/a	891	n/a	n/a

zone. The base saturation flow rate (3,600/base saturation headway) is then multiplied by the factor calculated in Table 5. The HCM methodology uses the calculated factor to adjust all lane groups on the approach. Since there was no after measured value for Site 5, no comparison with the other models could be made.

The Schroeder et al. (*1*) model was also used to predict saturation flow rate at each site, using the measured "base" saturation headway instead of the 2.04 base value shown in Equation (4). All the locations had low work intensity and normal pavement condition.

Table 6 shows the measured base saturation headway, the HCM estimated work zone saturation flow rate, the Schroeder et al. model estimated work zone saturation flow rate, and the saturation flow rate from field data during the work zone.

The capacity of the lane group is calculated using Equation (5). Table 7 shows the capacity per lane group and per approach from the field data, the HCM (*2*) estimated value, and the Schroeder et al. model (*1*). For all the sites, both methods underestimated the capacity of the approach.

$$c = s * N * g / C \tag{5}$$

where

c = capacity of lane group,
s = saturation flow rate per lane of lane group, vehicles per hour green per lane (vphgpl),
N = number of lanes in lane group during work zone, and
g/C = effective green to cycle length ratio.

Table 8 shows the percent difference of the two models from the field data. The HCM methodology underestimated the capacity of the approach when a work zone was present by from 9% to 23%. The Schroeder et al. methodology underestimated the capacity of the approach when a work zone was present by from 2% to 25%.

Conclusion

This research looked at a specific case study of vehicle behavior in work zones in NYC. Data was collected and compared to existing models to determine if any could reliably predict the effect of the work zone on the approach

Table 8. Percent Difference Comparison of Models versus Field Data

Site	Comparison field data and HCM, % difference	Comparison field data and schroeder et al. model, % difference
1	9%	17%
2	19%	18%
3	13%	18%
4	23%	25%
5	n/a	n/a

capacity. The data collected found that vehicle headways decreased in thru lanes during the period of time in which there was a work zone present. This was an unexpected result and the opposite result from what other models predict. The data was less conclusive for shared lanes. One reason for this was that in the shared lanes, it was more difficult to have cycles with queues of at least 6 to 7 vehicles with no heavy vehicles in the queue. The data did suggest that the lanes next to the work zone experienced no change in saturation headway. This is also different from the other models which predict all lanes will experience the same effect.

The model most commonly in use is the HCM model (*2*). The data used in calibrating this model was obtained at 6 intersections, none of which were in a CBD location. A second model created by Schroeder et al. (*1*) was developed after the publication of the HCM model. The Schroeder et al. model was calibrated from intersections in North Carolina that were also not located in CBDs. It was found that both models underestimate the capacity of the approach, sometimes by as much as 20%. This suggests that vehicles at intersections with short-term work zones in high-density CBD environments, such as NYC, behave differently than in work zones in non-CBD locations. A possible reason for this is that there is increased traffic pressure on drivers in a high-density CBD when a lane is closed. Traffic engineers would need to know this when planning for work zone lane closures.

Although the data for this research was limited, it is the conclusion of the authors that there is a strong suggestion that this behavior would be repeated at other CBD locations. It is the recommendation that further research is needed to more closely investigate the downward shift in mean headway at CBD locations under work zone conditions. Particularly, since the other models were calibrated based on data from intersections, none of which were in CBDs, more research is needed at CBD intersections. Additional data could then be used to calibrate and validate a model that would predict intersection saturation flow rate and capacity in such environments.

Author Contributions

The authors confirm contribution to the paper as follows: study conception and design: Nawaf Alshabibi, Elena Prassas; data collection: Nawaf Alshabibi; analysis and interpretation of results: Nawaf Alshabibi, Elena Prassas; draft manuscript preparation: Nawaf Alshabibi. All authors reviewed the results and approved the final version of the manuscript.

References

1. Schroeder, B., N. Rouphail, B. Williams, A. Hajbabaie, B. Narron, S. Kim, B. Aghdashi, and K. Tabrizi. *Delay and User Cost Estimation for Work Zones on Urban Arterials*. NCDOT Project 2013-09. Institute for Transportation Research and Education (ITRE), North Carolina State University, 2015.
2. *Highway Capacity Manual: A Guide for Multimodal Mobility Analysis,* 6th ed. Transportation Research Board, Washington, D.C., 2016.
3. Sarasua, W., W. Davis, D. Clarke, J. Kottapally, and P. Mulukutla. Evaluation of Interstate Highway Capacity for Short-Term Work Zone Lane Closures. *Transportation Research Record: Journal of the Transportation Research Board*, 2004. 1877: 85–94.
4. Al-Kaisy, A., M. Zhou, and F. Hall. New Insights into Freeway Capacity at Work Zones: Empirical Case Study. *Transportation Research Record: Journal of the Transportation Research Board*, 2000. 1710: 154–160.
5. Heaslip, K., A. Kondyli, D. Arguea, L. Elefteriadou, and F. Sullivan. Estimation of Freeway Work Zone Capacity through Simulation and Field Data. *Transportation Research Record: Journal of the Transportation Research Board*, 2009. 2130: 16–24.
6. Jiang, Y. Traffic Capacity, Speed, and Queue-Discharge Rate of Indiana's Four-Lane Freeway Work Zones. *Transportation Research Record: Journal of the Transportation Research Board*, 1999. 1657: 10–17.
7. Dixon, K., J. Hummer, and A. Lorscheider. Capacity for North Carolina Freeway Work Zones. *Transportation Research Record: Journal of the Transportation Research Board*, 1996. 1529: 27–34.
8. Bonneson, J., and K. Nguyen. *HCM Urban Streets Methodology Enhancements – Saturation Flow Rate Adjustment Factor for Work Zone Presence*. Working Paper No. 6, SHRP 2 Project LO8 Incorporation of Non-recurrent Congestion Factors into the Highway Capacity Manual Methods, 2012.
9. Hawkins, H., K. C. Kacir, and M. A. Ogden. *Traffic Control Guidelines for Urban Arterial Work Zones-Volume 2-Technical Report*. Texas Transportation Institute, FHWA/TX-91/1161-5, Vol. 2, 1992.
10. Joseph, C. T., E. Radwan, and N. M. Rouphail. Work Zone Analysis Model for the Signalized Arterial. *Transportation Research Record: Journal of the Transportation Research Board*, 1988. 1194: 112–119.
11. Elefteriadou, L., M. Jain, and K. Heaslip. *Impact of Lane Closures on Roadway Capacity*. Transportation Research Center, University of Florida, 2008 http://www.dot.state.fl.us/research-center/Completed_Proj/Summary_RD/FDOT_BD545_61_B.pdf.
12. Bonneson, J., B. Nevers, J. Zegeer, T. Nguyen, and T. Fong. *Guidelines for Quantifying the Influence of Area Type and Other Factors on Saturation Flow Rate*. Texas Transportation Institute, College Station, 2005. http://www.dot.state.fl.us/research-center/Completed_Proj/Summary_PL/FDOT_DO2319_rpt.pdf.
13. Li, H., and P. Prevedouros. Detailed Observations of Saturation Headways and Start-up Lost Times. *Transportation Research Record: Journal of the Transportation Research Board*, 2002. 1802: 44–53.

14. Hamad, K., and H. Abuhamda. Estimating Base Saturation Flow Rate for Signalized Intersections in Doha, Qatar. *Journal of Traffic and Logistics Engineering*, Vol. 3, No. 2, 2015, pp. 168–171.
15. Rahman, M. M., S. N. Ahmed, and T. Hassan. Comparison of Saturation Flow Rate at Signalized Intersections in Yokohama and Dhaka. *Proceedings of the Eastern Asia Society for Transportation Studies*, Vol. 5, 2005, pp. 959–966.
16. Majeed, A. A., S. O. Zephaniah, G. Mehta, and S. Jones. Field Based Saturation Headway Model for Planning Level Applications. *International Journal of Traffic and Transportation Engineering*, Vol. 3, No. 5, 2014, pp. 207–215. http://article.sapub.org/pdf/10.5923.j.ijtte.20140305.01.pdf.
17. Shang, H., Y. M. Zhang, and L. Fan. Heterogeneous Lanes' Saturation Flow Rates at Signalized Intersections. *Proc., 9th International Conference on Traffic and Transportation Studies, Proceedings – Social and Behavioral Sciences*, Science Direct, Procedia – Social and Behavioral Sciences 138 (2014) pp. 3–10 https://core.ac.uk/download/pdf/82096370.pdf

The Standing Committee on Highway Capacity and Quality of Service (AHB40) peer-reviewed this paper (18-03722).

Article

TRR
JOURNAL OF THE TRANSPORTATION RESEARCH BOARD

Transportation Research Record
2018, Vol. 2672(15) 9–22

DOI: 10.1177/0361198118788171
journals.sagepub.com/home/trr
SAGE

Total Capacity of Roundabouts Analyzed by a Conflict Technique

Ning Wu[1] and Werner Brilon[1]

Abstract

The conventional method of analysis for capacity of roundabouts is based on gap-acceptance methods or on empirical regression or a combination of both. The influence of pedestrians is modeled by reduction factors with a questionable empirical background. Thus, the current methods are based on an incoherent mix of approaches. Moreover, they do not account for the interaction between the different elements. The paper presents a new model which treats the whole intersection as one entity. Here all the conflict points where different streams (vehicles and non-motorized road users) intersect within the roundabout are identified. Each conflict point is treated as one queuing system with a simplified queuing mechanism. In addition, the interactions between the consecutive queuing systems are taken into account according to the theory of chains of queues where the distance (= storage area) between the conflict points becomes important. The paper explains the development of the model, presents the mathematical derivations of ready-to-use capacity equations plus parameter calibration by existing data, and demonstrates its real-world application. The advantage of the technique is that all conflicts—both between vehicle streams and with pedestrians at entries and exits—are treated by the same congruent methods. In addition, the interactions between consecutive arms of the roundabout are modeled. The method can also model limited priority (e.g., for pedestrians at crosswalks) for all conflict points. Finally, the conflicts that are decisive for the performance of the whole intersection are identified. The method can be easily implemented into computer software.

The capacity of roundabouts has been the subject of investigation in many countries since the implementation of modern roundabouts in the 1980s. The remarkable aspect is that each country has attempted to find its own solution. To some degree, this may be justified by different driver behaviors or divergent traffic rules. The differences between solutions in the various countries, however, have become rather small and may be mainly based on limited sample sizes or differing methods of analysis. Normally the classical approaches treat each entry to a roundabout as an isolated T-junction. The effects of mutual interaction between the various conflict points of the roundabout remain disregarded.

Based on gap-acceptance theory, the maximum throughput (= capacity, C) of one entry into the roundabout can be calculated by the Siegloch formula (*1*)

$$C = \frac{3600}{t_f} e^{-\frac{q_C}{3600} \cdot t_0} \tag{1}$$

where

C = capacity (veh/h),
$t_0 = t_c - t_f / 2$
= so-called zero-gap (s),
t_f = follow-up time (s),
t_c = critical gap (s), and
q_C = traffic volume on the circular lane in front of the entry (veh/h).

A consequent application would require reliable estimates for t_c and t_f (cf. [*2*]). This formula is derived based on the assumption that the gaps on the circular lane are exponentially distributed (i.e., no bunching). This is rather unrealistic at roundabouts. Also the effects of priority reversal or typical influences of the degree of saturation on t_c and t_f cannot be described.

Therefore, most researchers favor the so-called "empirical regression" approach for estimation of roundabout capacity. Here the real-world traffic at a roundabout entry is measured over periods of steady queuing on the entry during specific time intervals, for example, one minute. In the case of a steady queue on the entry lane, the observed flow volume is the capacity. This can

[1]Institute for Traffic Engineering and Management, Ruhr-University Bochum, Bochum, Germany

Corresponding Author:
Address correspondence to Werner Brilon: werner.brilon@rub.de

be plotted over the observed conflict volume q_C on the circular lane. A regression line is then applied to represent these results.

The earliest attempt at using this idea was the well-known British investigation by Kimber (*3*). Here a linear regression equation for C as a function of q_C was applied in which the parameters of the equation were modified according to the geometric features of the intersection. Linear equations were also applied elsewhere (e.g. [*4*]). Other authors found nonlinearities within their empirical results. Most of them used exponential regression functions. One such solution is to use t_c and t_f in Equation (1) as regression parameters (e.g. [*5*]). Others (e.g. [*6*]) transformed this into a simple exponential function. In this form, the capacity formulas of the 2016 *Highway Capacity Manual* (HCM) (*7*) have become a standard for the U.S.A.

Both the linear and exponential regression approaches for estimating the capacity at roundabouts are rather pragmatic. One cannot be sure that the linear or exponential functions do also apply in areas of the C–q_C diagram with few measurement points. Therefore, Wu (*8*) proposed the following, more general, formula for the capacity of a single-lane entry to a roundabout:

$$C = \underbrace{\frac{3600}{t_f}}_{T1} \cdot \underbrace{\left(1 - \frac{q_C \cdot \tau}{3600}\right)}_{T2} \cdot \underbrace{\exp\left(-\frac{q_C}{3600} \cdot (t_0 - \tau)\right)}_{T3} \quad (2)$$

where

C = capacity of the entry (veh/h),

q_C = flow on circular lanes at the subject entry (veh/h), and

τ = minimum headway between vehicles on the circular lane (s).

The specific advantage of this approach is that the minimum headway τ between successive vehicles on the circle is taken into account. Equation (2) is the generalized form for the exponential and the linear capacity function. For example, this formula converges to an exponential function by $\tau = 0$ and to a linear function by $t_0 = \tau$. The equation can be understood as a multiplicative combination of the three terms *T*1, *T*2, and *T*3.

There are also reports which investigate the influence of the exiting traffic on the entry capacity. These include Bovy et al. (*9*), for single-lane roundabouts in Switzerland, and Wu (*10*), for double-lane roundabouts with single-lane exits. Schmotz (*11*), for mini roundabouts, found its way into the German guidelines, *Handbuch für die Bemessung von Strassen verkehrsanlagen* (*Manual for the Dimensioning of Roads German Highway Capacity Manual*) (HBS) (*12*).

On the entry lanes to a roundabout, there may be another type of conflict requiring capacity consideration. Besides the conflicting flow of vehicles on the circular lane, this is the pedestrian crossing with priority over the entering motorized vehicles, for example, a zebra crossing. The capacity-reducing effect of pedestrian crossings on entry capacity has been studied by Stuwe (*13*). Her results, which were based on a rather limited sample size of German zebra crossings, have been implemented in the HBS (*12*) and also in the HCM (*7*).

In addition, exits from a roundabout can become a bottleneck for vehicle traffic when limited capacity is influenced by pedestrians crossing the exit. A solution to describe this capacity has been presented by Marlow and Maycock (*14*) in conjunction with Griffith's (*15*) capacity formula for vehicular traffic traversing a zebra crossing. Another more recent result for exit capacity at mini roundabouts was presented by Schmotz (*11*).

One commonality of all the results published up to now is that they treat specific conflicts at roundabouts with differing and incoherent methods and they do not consider any interaction between adjacent conflict areas. In this respect, the solutions may be realistic as long as each element of the intersection is not operating near or above capacity. As soon as one of the conflict points is overloaded for a short period, queues may form. Due to limited storage space on the circular lane, these queues impede other conflict points with the consequence of reduced capacities there. Thus, a short overload at one point can easily become the starting point for an overload of the whole intersection—even if the analysis for all the individual conflicts still gives a sufficient quality level of service. This kind of mutual interdependence of different parts of the roundabout is not confined to the case of a complete overload. Instead, it can occur with just a moderate increase of traffic demand. For this reason, it is desirable to formulate algorithms that describe the capacity of the roundabout as a whole, regarding the capacities of the various conflict points as well as the influence of limited internal storage areas.

This paper, like the preceding presentation (*16*), presents a mathematical model which is capable of estimating the capacity of the roundabout as a whole. The model represents the various conflict points by a homogeneous set of models including a stochastic approach to cope with the interlocking between subsequent bottlenecks within the intersection.

Methodology

Capacity of Traffic Streams with Priority Control

By conflict, we refer to the intersection of several streams that have to pass the same area within an intersection (cf. *17, 18*). The vehicles involved in a conflict have to pass the area one after the other. The set of streams involved in the same conflict is called a conflict group. The simplest case is a conflict group of two streams (Figure 1*a*). One of these streams (i) is assumed to have priority over the other (j).

It can be assumed that any major stream vehicle passing a conflict point will occupy that point for a certain time. Only during the time it is unoccupied can the minor stream pass the conflict point with a basic capacity C_0. The basic capacity of a minor stream is the reciprocal of the follow-up time t_f. The conflict point can be occupied by a major stream vehicle in three different ways: it is occupied if there is a queue, if a platoon is passing that point, or if a single vehicle is arriving. A minor stream vehicle can only pass the conflict point if the conflict point is neither occupied by a queue (spilling back from downstream) nor by a platoon nor by a single arriving vehicle in the major stream (*19*). The case of a spillback from downstream is treated in subsequent sections of this paper.

In cases in which no queue from downstream occurs in the major stream, the last two terms of Equation (2) describe exactly the portions of occupied time in the major stream. Thus, for a two-stream problem (Figure 1*a*), the capacity of a minor stream j entering a major stream i can be calculated by a general model closely related to Equation (2):

$$C_j = T1 \cdot T2 \cdot T3 \quad (3)$$

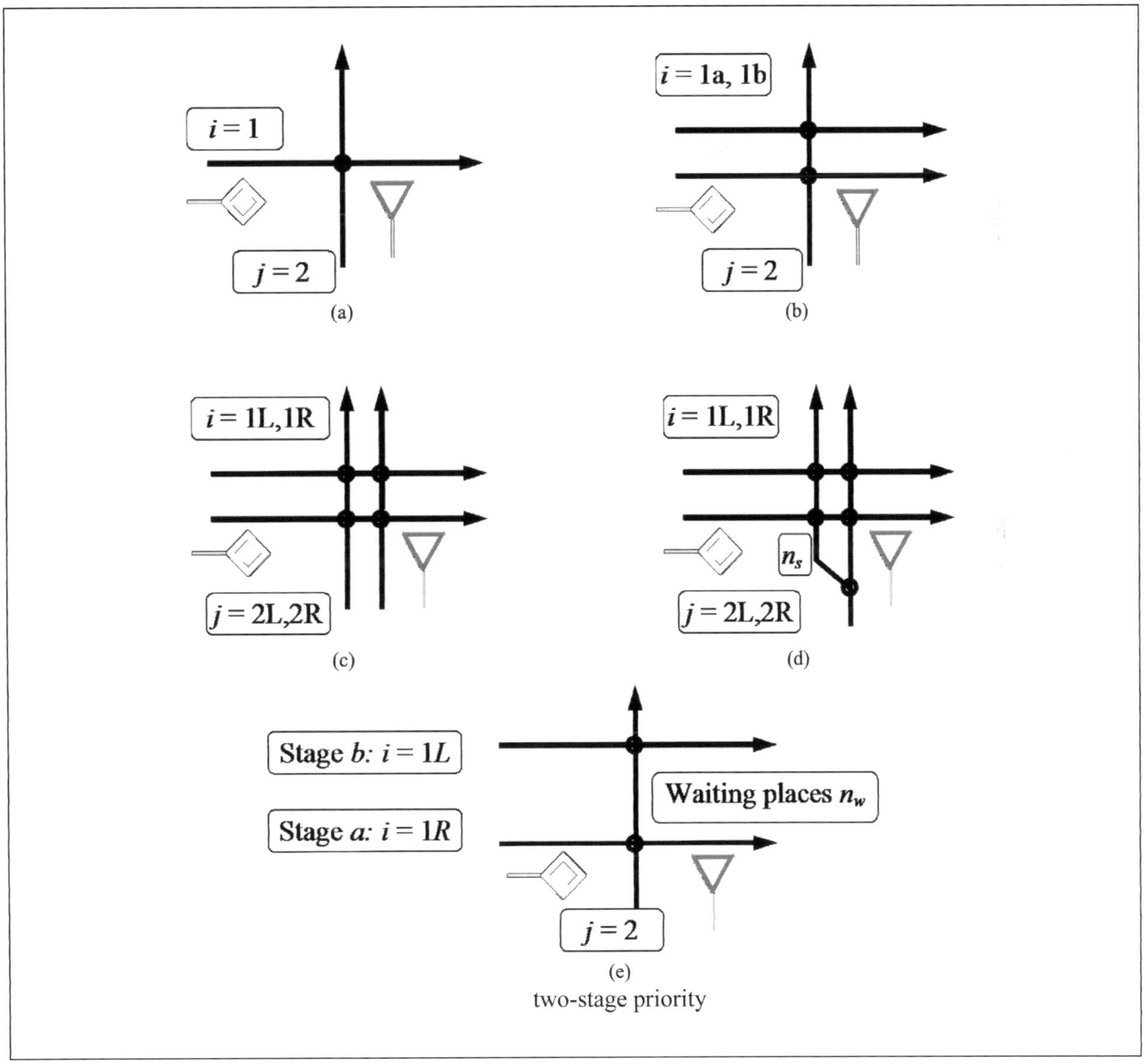

Figure 1. Conflict points at unsignalized intersections.

where

C_j = capacity of the minor stream j (veh/h),
$T1$ = $C_{0,j}$
= $3600/t_{f,j}$
= basic capacity of the minor stream j in case of no major stream (i) vehicle is occupying the conflict point (veh/h),
$T2$ = $1 - \frac{q_{C,i} \cdot \tau_i}{3600}$
= portion of time of no vehicle platooning (vehicles following each other consecutively with a minimum headway τ_i) in the major stream i, and
$T3$ = $\exp\left(-\frac{q_{C,i}}{3600} \cdot (t_{0,ij} - \tau_i)\right)$
= portion of time of no impedance caused by a vehicle arriving from upstream in the major stream i.

As a simplification, we can assume $t_{0,ij} \approx \tau_i$. This can be considered as realistic in most cases for traffic behavior at a roundabout due to the usual estimates for t_c and t_f. Thus, Equation (3) yields as an approximation for the capacity C_j:

$$C_j = C_{0,j} \cdot \left(1 - \frac{q_{C,i} \cdot \tau_i}{3600}\right) = \frac{3600}{t_{f,j}} \cdot \left(1 - \frac{q_{C,i} \cdot \tau_i}{3600}\right) \quad (4)$$

where

$t_{f,j}$ = follow-up time of the minor stream j (s),
$q_{C,i}$ = volume of the major stream near entry i (= circular stream at a roundabout) (veh/h), and
τ_i = minimum time headway of the major stream i (s).

For more than one major stream on the circular roadway (Figure 1*b*), the portion of time $T2$ of no vehicle platooning for different major streams i is assumed to be independent from each other. Thus, for a minor stream j with multiple major streams i, we obtain

$$C_j = C_{0,j} \prod_{i \in I} \left(1 - \frac{q_{C,i} \cdot \tau_i}{3600}\right) \quad (5)$$

where I is the set of all major streams under consideration.

Here also the so-called "limited priority" (cf. [*20*]) can be included. Limited priority means that the major stream vehicles do not use their right of way in each case. This effect can be taken into account by the probability b_{ij}. Equation (5) is then expanded into (cf. [*21*])

$$C_j = C_{0,j} \prod_{i \in I} \left(1 - b_{ij} \frac{q_{C,i} \cdot \tau_i}{3600}\right) \quad (6)$$

where

b_{ij} = probability that a major stream vehicle goes first in case of a conflict (e.g., b_{ij} = 1: no i yielding to j, or in other words: the priority of i is always obeyed by j). The values of b_{ij} can be defined according to a so-called conflict matrix.

The total capacity of an entry with more than one lane is the sum of the capacities of all entry lanes (Figure 1*c*). That is,

$$C = \sum_{j \in J} C_j = \sum_{j \in J} \left(C_{0,j} \prod_{i \in I} \left(1 - b_{ij} \frac{q_{C,i} \cdot \tau_i}{3600}\right)\right) \quad (7)$$

where J is the set of all minor streams under consideration.

Multilane Entry with Limited Length of Lanes

If an entry lane has only a limited length n_s, and both lanes emerge from a single upstream lane (Figure 1*d*), the capacities of the single lanes at the entry cannot be fully utilized as calculated. According to the usual concept of unsignalized intersections, this is a "short-share lane" problem. The capacity of shared lanes can be determined according to a formula first developed by Harders (*22*). This concept has been extended by Wu (*8*, *10*) such that additional lanes of limited length (short lane) can also be taken into account. For the case of a single-lane approach with an additional short lane near the intersection, the capacity of the entry with short-share traffic lane can be calculated (cf. [*8*, *10*]) from:

$$C_s = \frac{q_{2L} + q_{2R}}{\sqrt[n_s+1]{(x_{2L})^{n_s+1} + (x_{2R})^{n_s+1}}} = \frac{q_{2L} + q_{2R}}{\sqrt[n_s+1]{\left(\frac{q_{2L}}{C_{2L}}\right)^{n_s+1} + \left(\frac{q_{2R}}{C_{2R}}\right)^{n_s+1}}} \quad (8)$$

where

C_s = capacity of the short shared lane (veh/h),
q_{2L} = volume for the entry lane $2L$ (veh/h),
q_{2R} = volume for the entry lane $2R$ (veh/h),
C_{2L} = capacity for the entry lane $2L$ (veh/h),
C_{2R} = capacity for the entry lane $2R$ (veh/h),
x_{2L} = degree of saturation of the entry lane $2L$,
x_{2R} = degree of saturation of the entry lane $2R$, and
n_s = number of vehicles which can queue up on the short lane.

For n_s = 0, Equation (8) yields the well-known shared-lane formula from Harders (*22*):

$$C_s = \frac{q_{2L} + q_{2R}}{x_{2L} + x_{2R}} = \frac{q_{2L} + q_{2R}}{\frac{q_{2L}}{C_{2L}} + \frac{q_{2R}}{C_{2R}}} \quad (9)$$

The Two-stage Queuing Problem

Equation (6) applies only to cases in which the major streams are crossed by a minor stream at an isolated conflict point. In case of a pedestrian crossing at a roundabout, there are often several spaces between the circular major stream and the pedestrian crossing (Figure 1*e*). The minor stream at the entry can pass the pedestrian crossing and the circular major stream one by one, and if necessary wait in between. The capacity of this two-stage queuing system with n_w waiting places is given by Brilon and Wu (*23*):

$$C_T = \begin{cases} \frac{1}{n_w+1} \cdot (n_w \cdot C_b + C_{ab}) & \text{for } y = 1 \\ (1 - w_0) \cdot C_b + w_0 \cdot C_{ab} & \text{elsewhere} \end{cases} \quad (10)$$

where

C_T = total capacity of the combined queuing system (veh/h),

w_0 = $\dfrac{y-1}{y^{n_w+1}-1}$,

n_w = possible waiting places between both stages,

C_a = $C_{0,a}p_{0,a}$ = capacity of the first stage a (veh/h),

C_b = $C_{0,b}p_{0,b}$ = capacity of the second stage b (veh/h),

C_{ab} = $C_{0,ab}p_{0,a}p_{0,b}$ = capacity of queuing system in case $n_w = 0$ (veh/h),

$p_{0,a}$, $p_{0,b}$ = probabilities that the queuing stage a or b are not impeded by major vehicles, and

y = $\dfrac{C_a - C_{ab}}{C_b - C_{ab}}$

Different applications for the two-stage model at a roundabout are described in the subsequent sections of the paper. The calculation of C_a and C_b is explained for each case in these sections.

In Brilon and Wu (*23*), the capacity C_{ab} of the queuing system in case of $n_w = 0$ is only given for a special case with $C_{0,a} = C_{0,b} = C_{0,ab}$. For the two-stage problem at roundabouts regarding pedestrians, this assumption does not apply. However, in the general case with $C_{0,a} \neq C_{0,b} \neq C_{0,ab}$ the capacity C_{ab} can also be calculated. The capacity C_{ab} is a function of capacities of the two stages. That is,

$$c_{ab} = C_{0,ab}p_{0,a}p_{0,b} = C_{0,ab}\frac{C_a}{C_{0,a}}\frac{C_b}{C_{0,b}} = f_{ab}C_aC_b \quad (11)$$

where

$$f_{ab} = \frac{C_{0,ab}}{C_{0,a} \cdot C_{0,b}}$$

This leads to

$$y = \frac{C_a - C_{ab}}{C_b - C_{ab}} = \frac{C_a(1 - f_{ab}C_b)}{C_b(1 - f_{ab}C_a)} \quad (12)$$

The values of C_0 are actually the reciprocal of the follow-up time t_f. It is the time headway h at lane capacity C_{ln} plus the lost time Δt experienced by an approaching vehicle needed for deceleration and orientation maneuver. That is,

$$\Delta t = t_f - h = \frac{3600}{C_0} - \frac{3600}{C_{\text{ln}}} \quad (13)$$

where Δt is given in seconds.

For crossing stage a and b in one step, the deceleration and orientation time Δt is the sum of Δt in both stages. That is,

$$\Delta t_{ab} = \Delta t_a + \Delta t_b = \frac{3600}{C_{0,a}} + \frac{3600}{C_{0,b}} - \frac{2 \cdot 3600}{C_{\text{ln}}} \quad (14)$$

Thus,

$$C_{0,ab} = \frac{3600}{h + (\Delta t_a + \Delta t_b)} = \frac{1}{\dfrac{1}{C_{0,a}} + \dfrac{1}{C_{0,b}} - \dfrac{1}{C_{\text{ln}}}} \quad (15)$$

In the following, a time headway of h = 2.2 s is used at roundabouts. Thus, the lane capacity C_{ln} is 3600/2.2 s = 1640 veh/h.

Equation (10) was derived for a queuing system with Markovian arrivals and departures. In reality, those presumptions are rarely satisfied. To account for the stochastic property of a queuing system, a factor C_n can be applied to the parameter n_w. Thus we get as an approximation to the stochastic property of the queuing system instead of n_w a parameter $n_w^* = C_n n_w$ in Equation (10). For a queuing system with Markovian arrivals and departures $C_n = 1$. The factor C_n is normally larger than 1. For example, one can use $C_n = 1.68$ for a queuing system with Markovian arrivals and deterministic service times (*24*). The value of C_n is subject to calibration.

In general, the total capacity of a two-stage queuing system can then be expressed by four significant parameters:

$$C_T = f(C_a, C_b, C_{0,ab}, n_w^*) \quad (16)$$

The formulation (Equation (10)) of the capacity of a two-stage queuing system is very complex. As a simplification, the following formulation can be used instead of Equation (10):

$$C_T = \min\begin{cases} C_a \cdot \left(1 - (1 - f_{ab} \cdot C_b)^{n_w^*+1}\right) \\ C_b \cdot \left(1 - (1 - f_{ab} \cdot C_a)^{n_w^*+1}\right) \end{cases} = f(C_a, C_b, C_{0,ab}, n_w^*) \quad (17)$$

For practical applications, this approximation provides a rather good fit.

Conflict Groups at a Roundabout

All capacities of the individual conflicts or conflict groups at a roundabout can be estimated using the equations presented in the previous section. The total set of conflict points at a roundabout with four arms and 12 movements is illustrated in Figure 2, *a* and *b*. For a particular double-lane entry yielding to a double-lane circular roadway, the traffic streams and the involved conflict points are depicted in Figure 2*c*. The traffic lanes are named left (L) and right (R) entry lane, outer (O), inner (I) circular lane, and pedestrian crosswalk (PE).

The circular volume q_C, the entry volume q_E, and the exit volume q_A at each subject approach k can be easily calculated from the O–D matrix of traffic demands at the intersection.

For double-lane roundabouts, one can assume that the volume on the left entry lane (L) corresponds to the left-turn volume at the entry. Furthermore, it can be assumed that in practice none (0%) of the left-turn vehicles from the upstream arm will use the inner circular lane (I) at a very low circular volume and nearly all of them will use the inner circular lane at high circular volume (up to 1,600 veh/h). In between, the volumes can be obtained by interpolation. These assumptions are realistic because the inner circular lane and the left entry lane are mostly used by the corresponding left-turn vehicles. That is,

$$q_{C,I,k} = q_{L,k-1} \cdot \frac{q_C}{1600} = q_{k-1,k+2} \cdot \frac{q_C}{1600} \text{ and } q_{L,k} = q_{k,k+3} \tag{18}$$

For example, for an entry with two lanes (L and R), as shown in Figure 2*c*, with one outer (O), one inner (I) circular lane, and one pedestrian crosswalk (PE), the capacity of the left and the right entry are

$$C_L = C_{0,L} \cdot \left(1 - b_{IL} \cdot \frac{q_{c,I} \cdot \tau_I}{3600}\right) \cdot \left(1 - b_{OL} \cdot \frac{q_{c,O} \cdot \tau_O}{3600}\right) \cdot \left(1 - b_{PEL} \cdot \frac{q_{c,PE} \cdot \tau_{PE}}{3600}\right) \tag{19}$$

$$C_R = C_{0,R} \cdot \left(1 - b_{OR} \cdot \frac{q_{c,O} \cdot \tau_O}{3600}\right) \cdot \left(1 - b_{PER} \cdot \frac{q_{c,PE} \cdot \tau_{PE}}{3600}\right) \tag{20}$$

At roundabouts, there is usually a space with n_E storage places ($n_E \geq 1$) between the outer circular lane and the pedestrian crosswalk. A minor stream vehicle can cross the pedestrian crosswalk first and wait in between. In this case, the capacity of a minor stream must be calculated for crossing the pedestrian crosswalk (C_a) and for entering the major stream (C_b). Thus,

$$C_{L,a} = C_{0,PEL} \cdot \left(1 - b_{PEL} \cdot \frac{q_{C,PE} \cdot \tau_{PE}}{3600}\right) \tag{21}$$

$$C_{L,b} = C_{0,L} \cdot \left(1 - b_{IL} \cdot \frac{q_{C,I} \cdot \tau_I}{3600}\right) \cdot \left(1 - b_{OL} \cdot \frac{q_{C,O} \cdot \tau_O}{3600}\right) \tag{22}$$

The combined capacity of this two-stage problem can be obtained from Equation (10) or Equation (17):

$$C_{L,T} = f(C_{L,a}, C_{L,b}, C_{0,L,ab}, n_E) \tag{23}$$

Similarly, for the right entry lane we have

$$C_{R,a} = C_{0,PER} \cdot \left(1 - b_{PER} \cdot \frac{q_{c,PE} \cdot \tau_{PE}}{3600}\right) \tag{24}$$

$$C_{R,b} = C_{0,R} \cdot \left(1 - b_{OR} \cdot \frac{q_{c,O} \cdot \tau_O}{3600}\right) \tag{25}$$

$$C_{R,T} = f(C_{R,a}, C_{R,b}, C_{0,R,ab}, n_E) \tag{26}$$

where f is the functionality given by Equation (10) or Equation (17).

Obviously, for a multilane roundabout, the volume distribution between q_I and q_O on the circular lanes and the volume distribution between q_L and q_R on the entry lanes must be calculated in advance. The distributions of those volumes can be estimated according to the turning movements at the intersection. For a single-lane roundabout, for all equations the indices I and L are no longer applicable and terms containing these indices can be ignored.

Taking into account the volumes of both left (q_L) and right (q_R) entry lanes and the length of the double-lane area (n_d) upstream from the pedestrian crossing, the total capacity of the double-lane entry ($C_{E,d}$) can then be calculated using Equation (8).

$$C_{E,d} = \frac{q_L + q_R}{\sqrt[n_d+1]{\left(\frac{q_L}{C_{L,T}}\right)^{n_d+1} + \left(\frac{q_R}{C_{R,T}}\right)^{n_d+1}}} \tag{27}$$

Setting $n_d = 0$, this equation yields the shared lane capacity ($C_{E,s}$) of a single-lane entry opposing two circular lanes at a roundabout:

$$C_{E,s} = \frac{q_L + q_R}{\frac{q_L}{C_{L,T}} + \frac{q_R}{C_{R,T}}} \tag{28}$$

In this case two minor streams on one entry lane are actually calculated (cf. Figure 2*d*). One of them (q_L) crosses the outer circular lane and proceeds into the inner circular lane. And another (q_R) continues directly into the outer circular lane.

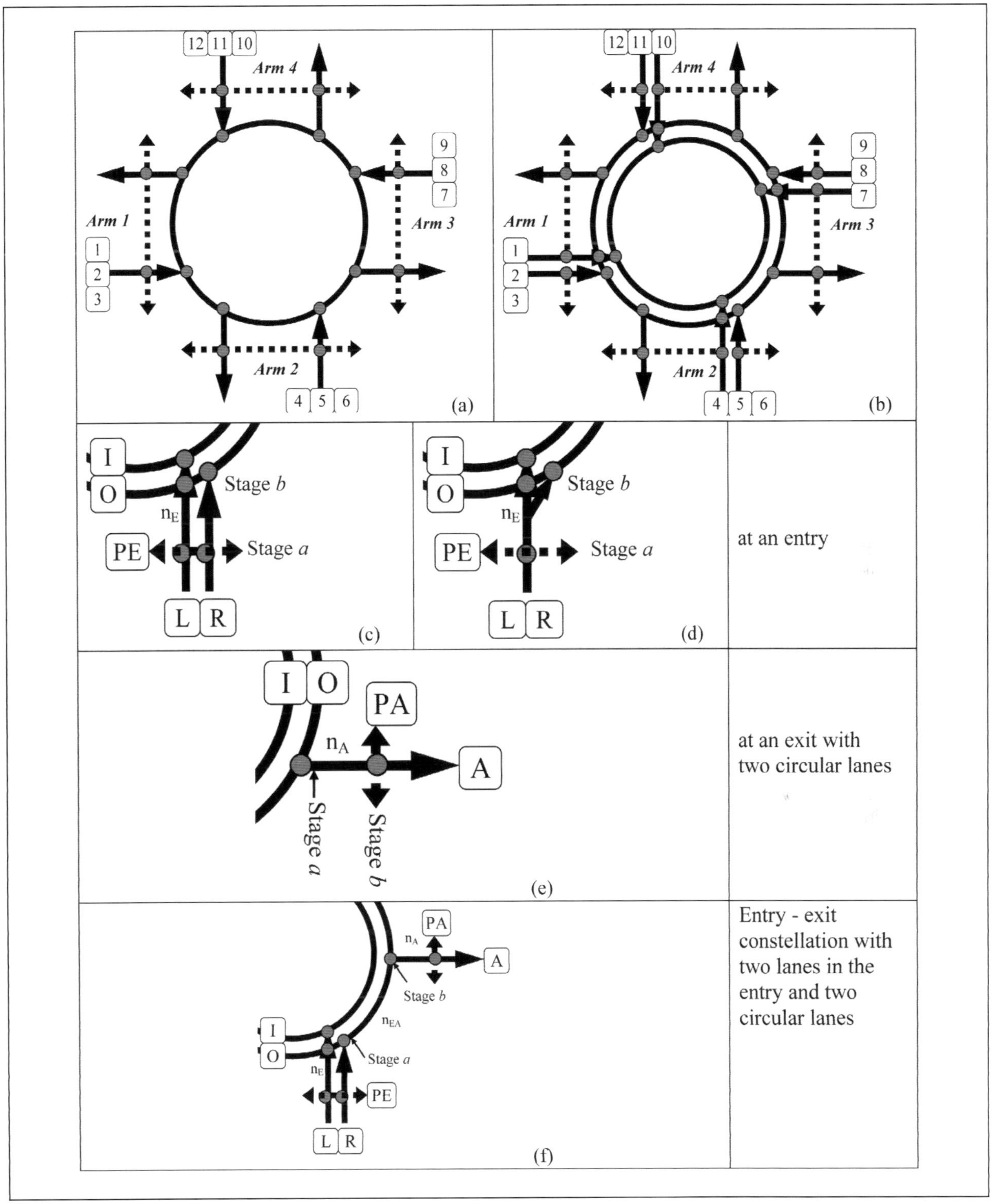

Figure 2. Conflict points at roundabouts.

Note that at a single-lane entry the approach arm normally has a flare area to enable the turning movement of the right-turn vehicle entering the roundabout. Thus, there is actually a double-lane area able to accommodate one vehicle. Under real-world conditions, this flare area is not used by all vehicles but only by a portion a_f of

Table 1. Conflict Matrix for Parameters of a Roundabout Entry for Different Configurations

Minor lane j	Major lane i	Circular inner (I) Stage b	Circular outer (O) Stage b	Ped (PE) Stage a	$C_{0,j}$ (veh/h) Veh Stage b	$C_{0,j}$ (veh/h) Ped Stage a	$C_{0,j}$ (veh/h) Two-stage ab	Places between two stages n_E (veh)
(a) Mini roundabout entry with single-lane entry and single circular lane (1/1 mini)								
	τ_i (s)	2.6		3.0				
Entry lane	b_{ij}	0.9		0.9	1080	1500	1020	1
(b) Roundabout entry with a single-lane entry and a single circular lane (1/1)								
	τ_i (s)	1.8 + 14.5/D*)		2.8				
Entry lane	b_{ij}	0.9		0.9	1200	1550	1150	1
(c) Roundabout entry with one entry lane and two circular lanes (1/2)								
	τ_i (s)	2.3	2.3	2.3				
Entry left (L)	b_{ij}	0.9	0.9	0.9	1270	1330	1080	1
Entry right (R)	b_{ij}	0.9	0.9	0.9	1420	1420	1250	1
(d) Roundabout entry with two entry lanes and two circular lanes (2/2)								
	τ_i (s)	2.4	2.4	2.6				
Entry left (L)	b_{ij}	0.9	0.9	0.9	1010	1010	730	1
Entry right (R)	b_{ij}	0.9	0.9	0.9	1100	1100	830	1
(e) Parameters of a roundabout exit with one exit lane								
	τ_i (s)	-		2.9				
Exit lane (A)	b_{ij}	1		0.9	1400	1550	1330	1
Upstream outer circular lane as a share lane (O,U)				Exit lane (A) Downstream outer circular (O,D)				$C_{A,T}$ 1640

(f) Parameters for an entry–exit constellation Minor lane j	Major lane i	Downstream outer lane at the entry with $C_j = C_{0,j}$ Stage a	Upstream outer lane at the exit with $C_j = C_{O,U}$ from Equation 34 Stage b	$C_{0,j}$ (veh/h) Stage a	$C_{0,j}$ (veh/h) Stage b	$C_{0,j}$ (veh/h) Two-stage ab	Places between two stages $C_n n_{EA}$ (veh)
	τ_i (s)	-	-				
Outer circular lane between an entry and an exit (EA)	b_{ij}	1	1	1640	1640	1640	1.68*3

them. Thus the capacity of the single-lane entry yielding to a double-lane circular roadway is

$$C_{E,\,single\ lane} = a_f C_{E,d} \cdot (n_d = 1) + (1 - a_f) \cdot C_{E,s}$$
$$= a_f \frac{q_L + q_R}{\sqrt[2]{\left(\frac{q_L}{C_{L,T}}\right)^2 + \left(\frac{q_R}{C_{R,T}}\right)^2}} + (1 - a_f) \frac{q_L + q_R}{\frac{q_L}{C_{L,T}} + \frac{q_R}{C_{R,T}}}$$

Again, the value of a_f is subject to calibration depending on the geometric layout of the entry area. With $a_f =$ 0 one is on the safer side of calculation.

Setting the traffic volume on the inner circular lane at $q_I = 0$, all formulas mentioned here can be used for a roundabout with a single-lane circular roadway.

In summary, all the parameters mentioned above can be defined using a conflict matrix. Table 1 shows parameters for the model calibrated to represent German roundabouts adjusted to methods from the HBS (*12*). The basic capacities of the minor stream $C_{0,j}$ and the minimum headway τ_I of the corresponding major stream mentioned above are given for traffic streams under consideration (veh, ped, and two-stage). For the calibration of double-lane roundabouts, it was assumed that in practice no (0%) circular vehicles use the inner circular lane at very low circular volume, and 30% at high circular volume (1,600 veh/h). For a double-lane entry, 30% of the total entry volume is assumed to use the left entry lane. These assumptions are realistic because the inner circular lane and the left entry lane are mostly used by the corresponding left-turn vehicles and due to the fact that in Germany drivers are rather reluctant to use the left lanes at a two-lane roundabout. That means:

$$\frac{q_{C,I}}{q_C} = \frac{0.3 \cdot q_C}{1600} \text{ or } q_{C,I} = \frac{0.3 \cdot q_C{}^2}{1600} \tag{30}$$

and

$$q_L = 0.3 \cdot q_E \tag{31}$$

In Figure 3, *a* and *b* a comparison of capacities obtained from the model compared with the HBS *(12)* formulas is depicted. It is obvious that the simplified model matches the HBS data very well.

Figure 3, *c–h* show a comparison of the model with the HBS method regarding the pedestrian impedance factor at roundabout entries. For single-lane roundabouts the model results represent the HBS methodology quite well (Figure 3, *c* and *d*). For double-lane roundabouts, the model results cannot represent the HBS method for the whole range of circular volumes. The reason for these deviations must be found in the HBS model, because the HBS model is a regression based on very limited data without any theoretical background. However, in the common range of circular volumes the results of HBS data can also be properly represented by the new model.

Capacity of an Exit from a Roundabout

An exit from a roundabout can also be considered as a two-stage queuing system (cf. Figure 2*e*). The combined capacity of this two-stage problem can be obtained from Equation (10) or Equation (17):

$$C_{A,T} = f(C_{A,a}, C_{A,b}, C_{0,A,ab}, n_A) \tag{32}$$

where

$C_{A,a}$ = C_{ln}
= capacity of the exit lane at the edge of the outer circular lane,

$C_{A,b}$ = capacity of the pedestrian crossing at the exit, and

$C_{0,A,ab}$ = the basic capacity of the two-stage problem at the exit.

The capacity $C_{A,b}$ can be calculated as

$$C_{A,b} = C_{A,PA} = C_{0,RAR} \cdot \left(1 - b_{PAR} \cdot \frac{q_{a,PA} \cdot \tau_{PA}}{3600}\right) \tag{33}$$

The upstream outer circular lane at an exit can be treated as a shared lane consisting of the exit lane (q_A) and the downstream outer circular lane ($q_{O,D}$). Thus, the capacity of the upstream outer circular lane at an exit is

$$C_{O,U} = \frac{q_A + q_{O,D}}{\frac{q_A}{C_{A,T}} + \frac{q_{O,D}}{C_{O,D}}} \tag{34}$$

The parameters for an exit are given in Table 1 (line e). The lane capacity of the downstream outer circular lane $C_{O,D}$ is set to the lane capacity C_{ln} (1,640 veh/h).

Reduction of Entry Capacity at a Roundabout Due to Queuing at the Downstream Exit

For both the left and right entry lanes, capacity can be impeded by vehicles queuing on the outer circular lane caused by delay at the downstream exit. Assuming that the downstream exit obeys an M/G/1 queuing system, the impedance factor $f_{imp,EA}$ can be calculated approximately as follows.

$$f_{imp,EA} = p_{0,EA,n_{EA}} = 1 - x_{O,EA}{}^{C_n \cdot n_{EA} + 1} \tag{35}$$

where

$p_{0,EA,n_{EA}}$ = probability that the entry is not impeded by queuing from the downstream exit,
= $\Pr(Q_{O,EA} \leq n_{EA})$
= probability that the space between the downstream exit and the subject entry is not totally occupied by the downstream queue,

$x_{O,EA}$ = $q_{O,EA}/C_{O,EA}$
= degree of saturation of the downstream outer circular lane,

$q_{O,EA}$ = volume of the downstream outer circular lane (veh/h),

$C_{O,EA}$ = capacity of the downstream outer circular lane (veh/h),

n_{EA} = storage places between the entry and the downstream exit on the outer circular lane (veh), and

C_n = factor accounting for the stochastic property of the queuing system.

In this equation, the service time of the queuing system is considered as less stochastic using a factor C_n = 1.68 applied to the parameter n_{EA} assuming the service time of the queuing system on a circular lane is nearly deterministic (*24*). In Table 1 (line f), the parameters for calculating the impedance caused by the downstream queue are given. The value of n_{EA} = 3 is assumed for a midsize single-lane roundabout with an outer diameter D = 35m. For a real-world roundabout, the value of n_{EA} can be obtained from the given geometry.

Considering the impedance of queuing caused by the downstream exit, the capacity of the second stage of the left lane is

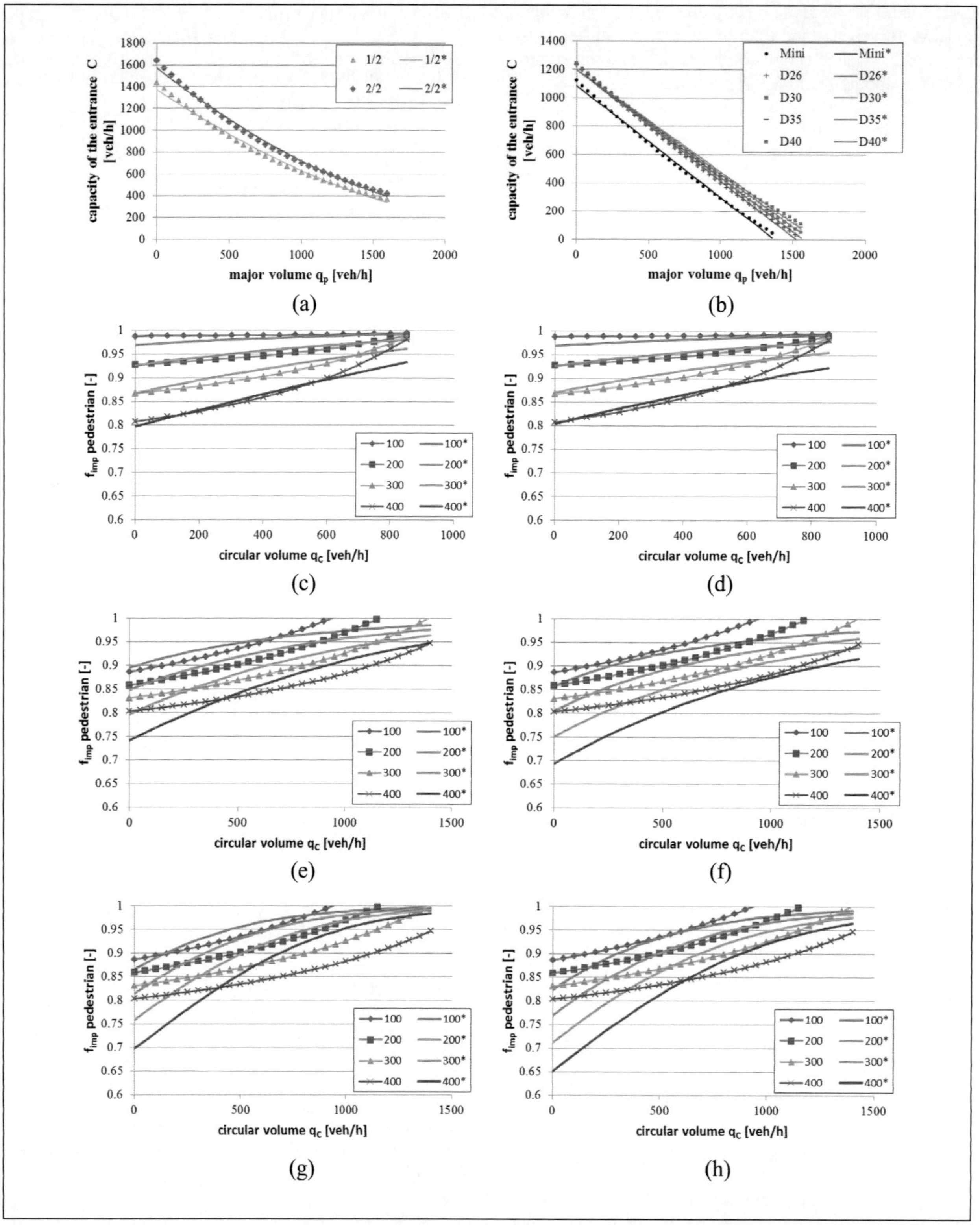

Figure 3. Comparison of the model calibration with HBS 2015 data (*12*) for basic entry capacity (*a*) and (*b*), and for the impedance factor for pedestrians: (*c*) mini, (*d*) 1/1, (*e*) 1/2L, (*f*) 1/2R, (*g*) 2/2 L, (*h*) 2/2R. dots = HBS, lines = new model.

$$C_{L,b}{}^* = \prod p_{0,q,i} \cdot C_{L,b} = p_{0,EA,n_{EA}} \cdot C_{L,b} = f_{imp,EA} \cdot C_{L,b} \tag{36}$$

with $C_{L,b}$ from Equation (22) and $f_{imp,EA}$ from Equation (35).

The capacity of the first stage remains unchanged:

$$C_{L,a}{}^* = C_{L,a}$$

with $C_{L,a}$ from Equation (21) and the functionality f from Equation (10) or Equation (17),

$$C_{L,T}{}^* = f(C_{L,a}{}^*, C_{L,b}{}^*, C_{0,L,ab}, n_E) \tag{37}$$

Similarly, for the right entry lane we get

$$C_{R,b}{}^* = p_{0,EA,n_{EA}} \cdot C_{R,b} = f_{imp,EA} \cdot C_{R,b} \tag{38}$$

$$C_{R,a}{}^* = C_{R,a} \tag{39}$$

with $C_{R,a}$ from Equation (24) and $C_{R,b}$ from Equation (25).

Thus, with functionality f from Equation (10) or Equation (17),

$$C_{R,T}{}^* = f(C_{R,a}{}^*, C_{R,b}{}^*, C_{0,R,ab}, n_E) \tag{40}$$

with all parameters defined previously.

For applications in practice, the following steps of an algorithm with the corresponding equations are summarized as a guide for the calculation procedure at an entry–exit constellation:

1 Estimation of the demand volumes q_C, q_E, and q_A at the subject entry and exit from the O–D matrix of traffic demand at the intersection
2 Estimation of the distribution of demand volumes by lanes (where applicable) both for circular lanes q_I and q_O and entry lanes q_L and q_R according to applicable assumptions
3 Downstream exit:
   3.1 Estimation of the first stage capacity C_a of the exit lane at the edge of the outer circular lane (e.g. C_a = 1,400 veh/h)
   3.2 Estimation of the second stage capacity C_b against the pedestrian stream (C_b from Equation (33))
   3.3 Estimation of the two-stage capacity $C_{A,T}$ at the exit lane ($C_{A,T}$ from Equation (32), C_a from point 3 and C_b from point 3.2)
   3.4 Estimation of the shared-lane capacity $C_{O,U}$ of the diverge point at the exit (Equation (34)) consisting of the downstream circular lane ($C_{O,D}$ = C_{In}) and the exit lane ($C_{T,A}$ from point 3.3)
   3.5 Estimation of the impedance factor $f_{imp,EA}$ (Equation (35)) to the upstream entry lanes due to queues from the downstream circular lane ($C_{O,U}$ from point 3.4)
4 Upstream entry (entry under consideration):
   4.1 Estimation of the first stage capacities C_a on entry lanes against the pedestrian stream (Equations (21) and (24))
   4.2 Estimation of the second stage capacities C_b on entry lanes against the circular major stream (Equations 22 and 25)
   4.3 Estimation of the second stage capacities C_a^* and C_b^* impeded by the downstream impedance factor (Equations 36 and 38, $f_{imp,EA}$ from section 3.5)
   4.4 Estimation the two-stage capacities $C_{L,T}^*$ and $C_{R,T}^*$ at the entry lanes (Equations 37 and 40)
   4.5 Estimation of the shared lane capacity $C_{E,d}$ (if applicable, Equation (27), with $C_{L,T}$ = $C_{L,T}^*$ and $C_{R,T}$ = $C_{R,T}^*$ from point 4.4)

Evaluation of the Proposed Model

To evaluate the proposed model, several examples have been constructed for a single-lane roundabout using the calibrated parameters in the previous sections. One example of an entry–exit constellation with blockage effect due to queue spillback from the downstream exit is demonstrated in Wu and Brilon (*16*).

In the second example, we consider a whole single-lane roundabout with all the entries and exits in a consecutive way. Here the queues from an entry will impede the upstream exit and a downstream exit may impede the upstream entry. This interrelation happens between all of the four arms of the roundabout. This circular interference requires an iterative calculation.

For the traffic demand, we consider that the roundabout is connecting a major (arms 1 and 3, cf. Figure 2*a*) and a minor (arms 2 and 4) road. The flow split of the major and the minor volume is 60% to 40% of the total intersection volume. The volumes of the turning movements are defined as 20%/60%/20% at a major street arm and 30%/40%/30% at a minor street arm (right/through/left).

We assume that one vehicle (n_{AE} = 1) can be stored on the circle between the exit and the entry (at the same arm) and three vehicles can be stored between an entry and the subsequent exit (n_{AE} = 3).

The roundabout can be preloaded by any smaller total traffic volume (e.g. 1,000 veh/h). Then the volume can be increased stepwise and the degrees of saturation of all the queuing systems on the roundabout can be evaluated iteratively counterclockwise at each step. The iteration is ended when the iteration will no longer lead to a stable solution. That would mean that, with an increase of

Table 2. Capacity of the Conflict Points on the Circular Lane Considering all Arms with Circular Queuing Interference at a Total Volume of 1,918 veh/h

		On circular lane						At corresponding arm		
				Downstream section						
Cross section at arm k		q	C^*	n_{+1}	$f_{imp,+1}$	C	x	q	C	x
1	Exit	959	1347	$n_{AE}=1$	0.96	1296	0.74	575	1382	0.42
	Entry	384	1640	$n_{EA}=3$	0.79	1298	0.30	575	811	0.71
2	Exit	959	1377	$n_{AE}=1$	0.90	1243	0.77	384	1382	0.28
	Entry	575	1640	$n_{EA}=3$	0.84	1373	0.42	384	936	0.41
3	Exit	959	1347	$n_{AE}=1$	0.96	1296	0.74	575	1382	0.42
	Entry	384	1640	$n_{EA}=3$	0.79	1298	0.30	575	811	0.71
4	Exit	959	1377	$n_{AE}=1$	0.90	1243	0.77	384	1382	0.28
	Entry	575	1640	$n_{EA}=3$	0.84	1373	0.42	384	936	0.41
1	Exit	959	1347	$n_{AE}=1$	0.96	1296	0.74			
						max x	0.77			0.71

q = traffic volume at the considered cross section on the circular lane (veh/h).
n_{+1} = storage places on the subsequent downstream section of the circular lane (veh).
= n_{EA} (storage places between the entry and the downstream exit on the circular lane).
= n_{AE} (storage places between the exit and the entry on the circular lane).
C^* = capacity of the considered conflict point without impedance of downstream queuing (veh/h).
C = $C^* f_{imp+1}$.
= capacity of the considered conflict point with impedance by downstream queuing (veh/h).
f_{imp+1} = impedance factor of downstream queuing on the subsequent section of the circular lane (cf. Equation 47).
(e.g., for the cross section Exit 3 is $C^* = 1347$, $f_{imp+1} = 1 - 0.3^{(1.68\times 1+1)} = 0.96$, $C = 1347^* \times 0.96 = 1296$).

demand, the capacity is going to be reduced due to spill-back effects on the circle.

In the example this point is reached at a total demand volume of 1,918 veh/h (= critical total volume). The conditions at this point are characterized by Table 2. Beyond this volume, the roundabout is expected to become grid-locked and a breakdown will occur with no chance of recovery as long as the demand volumes remain constant. During the iteration the gridlock effect becomes obvious by a steady increase of the x-values and a steady reduction of the capacities (down to zero beyond the critical volume) during the run of the iteration. At the critical volume, the maximum degree of saturation on the circular lane in the example is only 0.77. The corresponding maximum degree of saturation at the approach entries and exits is 0.71 (without regard to impedance on the circular lane). The total capacity of 1,918 veh/h contrasts to 2,720 veh/h which would result as the total capacity from the HBS methods without regard to queue interference effects.

Studying further examples it becomes obvious that the split of demand between major and minor streets as well as the proportions of the turning movements have an influence on the maximum total intersection capacity. For details, see Wu and Brilon (*16*). It was found that a misbalanced split of demands would cause a reduction of the total intersection capacity. Analogously, the total intersection capacities of double-lane roundabouts can be estimated as well.

Conclusion

This paper presents a model for capacity analysis of roundabouts with a completely new approach. It is based on the analysis of conflicts within the roundabout. The conflict points between the traffic streams of different types (cars and pedestrians) are considered using a homogenous model. The interaction between consecutive conflict points can be modeled according to the impedance probabilities. Together with this, the distance between the consecutive conflict points is modeled properly.

Based on the proposed model, the capacity of the total roundabout can be analyzed more precisely. In particular, the interaction between different traffic streams and consecutive conflict points at roundabouts can be accurately taken into account according to the model. Using the proposed model, the whole roundabout can be treated as one entity and the total intersection capacity can be obtained according to given traffic volumes for the movements at the intersection. As a result, the capacities of all the conflict points together with their degrees of saturation are obtained.

As one important result, it becomes obvious that the interference of potential queuing processes between conflicts on the circular lanes cannot be neglected, as is the case for all the conventional roundabout capacity calculation methods. The current practice may lead to a significant overestimation of the total intersection capacity. Above a degree of saturation of $x = 0.7$ (obtained by conventional capacity estimation) a risk of queuing gridlock on the circle may occur.

The correct application of the derived equations may be rather complex. It is, however, not too problematic to implement them into a computer program. To transfer the model to other countries a recalibration of the decisive parameters may be useful. The model has to be modified for multilane roundabouts where the exiting traffic is interfering with traffic from the upstream entry.

Further research may be directed at the influence of the b_{ij} (degrees of priority observation) between vehicles and in vehicle/pedestrian conflicts. Additional empirical research should also analyze how these b_{ij} are dependent on the saturation of the intersection. In addition, an empirical verification of the whole model will be a task for future research.

Author Contributions

The authors confirm contribution to the paper as follows: study conception and design, analysis and interpretation of results, and draft manuscript preparation performed by both authors. Both authors reviewed the results and approved the final version of the manuscript.

References

1. Siegloch, W. Die Leistungsermittlung an Knotenpunkten ohne Lichtsignalsteuerung (Capacity Calculations for Unsignalized Intersections). *Series Strassenbau und Strassenverkehrstechnik*, Vol. 154, 1973.
2. Brilon, W. Some Remarks Regarding the Estimation of Critical Gaps. *Transportation Research Record: Journal of the Transportation Research Board*, 2016. 2553: 10–19.
3. Kimber, R. M. *The Traffic Capacity of Roundabouts*. TRRL Laboratory Report 942. Transport and Road Research Laboratory (TRRL), Crowthorne, United Kingdom, 1980.
4. Brilon, W., L. Bondzio, and N. Wu. Unsignalized Intersections in Germany–A State of the Art. *Proc., 3rd International Symposium on Intersections Without Traffic Signals*. TR Circular E-C018. Transportation Research Board, Portland, Oregon, 1997, pp. 61–70.
5. Brilon, W., and N. Wu. Kapazitaet von Kreisverkehren – Aktualisierung (An Update for the Capacity of Roundabouts). *Strassenverkehrstechnik*, Nr. 5, 2008, pp. 280–288.
6. Rodegerdts, L. A., A. Malinge, P. S. Marnell, S. G. Baird, M. J. Kittelson, and Y. S. Mereszcak. *Accelerating Roundabout Implementation in the United States, Volume 2: Assessment of Roundabout Capacity Models for the Highway Capacity Manual*. Publication No. FHWA-SA-15-070. Federal Highway Administration United States Department of Transportation, Washington, D.C., 2015.
7. *Highway Capacity Manual: A Guide for Multimodal Mobility Analysis*, 6th ed. Transportation Research Board, Washington, D.C., 2016.
8. Wu, N. An Universal Formula for Calculating Capacity at Roundabouts. *Arbeitsblaetter, Institute for Traffic Engineering, Ruhr-University Bochum*, No. 13, March 1997.
9. Bovy, H., K. Dietrich, and A. Harmann. *Guide Suisse des Giratoires (Swiss Guideline for Roundabouts)*, Lausanne, February 1991, S. 75.
10. Wu, N. Capacity Enhancement and Limitation at Roundabouts with Double-lane or Flared Entries. *Proc., 5th International Symposium on Highway Capacity and Quality of Service*. Yokohama, Japan, July 25–29, 2006.
11. Schmotz, M. *Bemessungsverfahren für Minikreisverkehre und einstreifige Kreisverkehre (Design of Mini-roundabouts and Single-lane Roundabouts)*. Technical University of Dresden, 2014.
12. HBS 2015. *Handbuch für die Bemessung von Strassen* (FGSV, eds.), Research Association for Highways and Traffic, Cologne, 2015.
13. Stuwe, B. *Untersuchung der Leistungsfaehigkeit und Verkehrssicherheit an deutschen Kreisverkehrsplaetzen (Capacity and Traffic Safety at German Roundabouts)*. Ruhr-University Bochum, Institute for Transportation, No. 10, 1992.
14. Marlow, M., and G. Maycock. *The Effect of Zebra Crossing on Junction Entry Capacities*. TRRL Supplementary Report 724. Transport and Road Research Laboratory (TRRL), Crowthorne, United Kingdom, 1982.
15. Griffith, J. D. A Mathematical Model of a Nonsignalized Pedestrian Crossing. *Transportation Science*, Vol. 15, No. 3, 1981, pp. 222–232.
16. Wu, N., and W. Brilon. Roundabout Capacity Based on Conflict Technique. Presented at the 5th International Conference on Roundabouts, Green Bay, Wisconsin, May 2017, http://teachamerica.com/RAB17/RAB17papers/RAB175C_BrilonPaper.pdf.
17. Wu, N. Determination of Capacity at All-Way Stop-Controlled (AWSC) Intersections. *Transportation Research Record: Journal of the Transportation Research Board*, 2000. 1710: 205–214.
18. Brilon, W., and N. Wu. Unsignalized Intersections—A Third Method for Analysis. *Proc., 15th International Symposium on Transportation and Traffic Theory* (Taylor, A. P. ed.), Pergamon, Elsevier Science Ltd., New York, Tokyo, Oxford, 2002.
19. Wu, N. A Universal Procedure for Capacity Determination at Unsignalized (Priority-controlled) Intersections. *Transportation Research B*, Vol. 35, No. 3, 2001, pp. 592–623.
20. Troutbeck, R. J., and S. Kako. Limited Priority Merge at Unsignalized Intersections. Transportation Research Part A: Policy and Practice, Vol. 33, No. 3–4, 1999, pp. 291–304.
21. Miltner, T. *Verkehrsqualitaet an vorfahrtgeregelten Innerortsknotenpunkten (Traffic Flow Performance at Unsignalized Urban Intersections)*. Ruhr-University Bochum, Institute for Transportation, No. 27, 2003.
22. Harders, J. Die Leistungsfaehigkeit nicht signalgeregelten staedtischen Verkehrsknoten. (The Capacity of Unsignalized Urban Intersections) *Series Strassenbau und Strassenverkehrstechnik*, Nr. 76, 1968.

23. Brilon, W., and N. Wu. Capacity at Unsignalized Two-Stage Priority Intersections. *Transportation Research*, A, Vol. 33, No. 3-4, 1999.
24. Wu, N. An Approximation for the Distribution of Queue Lengths at Unsignalized Intersections. In *Proc., Second International Symposium on Highway Capacity, Sydney, 1994* (Akcelik, R. ed.), Australian Road Research Board Ltd., Victoria, Australia, 1994.

The Standing Committee on Highway Capacity and Quality of Service (AHB40) peer-reviewed this paper (18-00849).

Article

TRR
JOURNAL OF THE TRANSPORTATION RESEARCH BOARD

Transportation Research Record
2018, Vol. 2672(15) 23–34

DOI: 10.1177/0361198118794287
journals.sagepub.com/home/trr
SAGE

Evaluation of the Roundabout Capacity Model in HCM6 Edition and HCM 2010 on a Multilane Roundabout

Rongsheng Chen[1] and John Hourdos[1]

Abstract

This study evaluated the performance of the new Highway Capacity Manual 6th Edition (HCM6) roundabout capacity model at a multilane roundabout in Richfield, Minnesota. Traffic flow rates and gap acceptance data were collected during 20 hours of afternoon peak period traffic over four days. The observed critical headways for the left and right lanes was equal to 4.43 and 3.99 s while the observed follow-up headways were 3.05 and 2.96 s, respectively. Roundabout capacity curves for both lanes were built through the collected flow information, as well as through the observed headways and the HCM formula. As compared with the field observed results, the default model in HCM6 overestimated the capacity of the study roundabout. The default model in HCM 2010 closely estimated the capacity for the right lane, but overestimated the capacity of the left lane when the circulating flow rate was high. The HCM6 model was calibrated with the observed critical and follow-up headways, as well as by calibrating the intercept of the model using only the follow-up headway. The fully calibrated model overestimated the capacity of the right lane by 7–10% as the circulating flow decreases and overestimated the capacity of the left lane by 7% and 31%. The partially calibrated model overestimated the capacity of the left lane by 6% and overestimated the capacity of the right lane by at most 10% under low circulating flow rate, but it underestimated the capacity by at most 21% under high circulating flow rate.

Roundabouts are increasingly gaining popularity. It has been shown conclusively that roundabouts effectively enhance the safety of an intersection by greatly reducing fatal and severe injury crashes. This is achieved by reducing the number of conflict points, as well as the travel speed. Introducing any new facility into the road network, especially one as fundamentally different as a roundabout, requires accurate prediction of its capacity. Traditional traffic simulation models, although they can be calibrated to emulate a roundabout of a given capacity, are not robust enough to produce this from general driver behavior information. Specialized applications such as SIDRA (*1*), RODEL (*2*), and TORUS (*3*) are available, but introduce extra effort into the modeling process.

In the US, as in the case with many other road designs, the Highway Capacity Manual (HCM) has been the most used source for capacity and Level of Service (LOS) estimation and prediction. HCM models are based on empirical information and models developed from actual observations. Given the relatively young age of roundabouts as a road facility, the HCM roundabout model is still in flux. The latest HCM 6th Edition (HCM6) (*4*), has revised the proposed capacity model, citing observations of capacities higher than the ones predicted by the HCM 2010 (*5*). Although the HCM suggest local calibration of its model through gap acceptance information from similar roundabouts in the area, most users just use the default model. All HCM models reflect an average condition based on observations from several US roundabouts. Given the low number of roundabouts over the variety of design features, as well as differences in driver population, the average may not truly reflect the local conditions. A similar problem can affect the model calibrated as suggested in the HCM6, as it is only based on one driver behavior parameter, assuming all else remains the same.

This paper compares the capacities produced by the relevant HCM6 and HCM 2010 models with actual observations from a two-by-two (two-lane entrances/two-lane exits) roundabout in Richfield, MN built in

[1]Department of Civil, Environmental, and Geo-Engineering, University of Minnesota, Minneapolis, MN

Corresponding Author:
Address correspondence to Rongsheng Chen: chen4416@umn.edu

2008 (*6*). A separate, partially funded by NSF through grant #CNS-1544887, project which developed new sensing technologies installed a permanent high-resolution camera capable of observing the entire roundabout. The evaluation is based on direct measurements of flow, as well as on observations of critical and follow-up headways collected from 5 hours of video on each of the four entrances of the roundabout.

Background

HCM 2000 was the first version that included a capacity model covering only single-lane roundabouts. HCM 2010 simplified the form of the capacity model, as well as introducing variations covering multi-lane roundabouts. The recently released HCM6 retained the HCM 2010 form of the capacity model but updated the values of its parameters. These models, given the traffic flow rates of entering and circulating vehicles, estimate the roundabout capacity for the given entrance. The HCM 2010 models were based on 2003 studies of existing roundabouts conducted as part of NCHRP Project 3-65. A full description of the results of this project can be found in NCHRP Report 572 (*7*). HCM6 builds on the earlier models but is based on new data collected in 2012 as part of a FHWA supported project, presented in "Accelerating Roundabout Implementation in the United States – Volume II" report (*8*). In this study, a total of 819 usable minutes of data at single-lane roundabouts, 711 minutes of usable data for the multilane case for two entry lanes conflicting with two circulating lanes, and 519 minutes of usable data for the multilane case for two entry lanes conflicting with one circulating lane were obtained. The new data showed that the HCM 2010 generally underestimated capacity. For the HCM6, a recalibration of the HCM 2010 models was performed with an intercept that is calibrated to the follow-up time and a slope parameter based on the regression of the data. Additional calibration factors such as critical headway or geometric parameters beyond the number of lanes were not found to add significant improvement.

Most of the current evaluation studies were focused on comparisons with the NCHRP Report 572. Xu and Tian measured the critical and follow-up headways at ten roundabouts in California (*9*). They found that the critical headway in California was within the range reported in the NCHRP Report 572, but the follow-up headway was smaller.

Zheng et al. collected the critical and follow-up headways at four roundabout approaches in Wisconsin (*10*). Both headways for the single lane roundabout were within the range reported in the NCHRP Report 572. The measured critical headways for the right lane and left lane of a two-by-two roundabout were lower than the average values in the NCHRP Report 572. The measured follow-up headway for the same roundabout were similar to the values in the NCHRP Report 572.

Gazzarri et al. compared the gap acceptance parameters at seven roundabouts in Italy with the parameters in HCM 2010 (*11*). Three of the roundabouts were single-lane and the rest two-by-twos. They found that the measured critical and follow-up headways were smaller than that in HCM 2010.

Fitzpatrick et al. discussed the relationship between the temporal critical gap and the spatial critical gap and estimated the critical gap at roundabouts in the University of Massachusetts (*12*). The measured temporal critical headway was 2.2 s, which was significantly smaller than that stated in the HCM 2010.

Wei and Grenard calibrated the capacity model in HCM 2010 using gap-acceptance data from three single-lane roundabouts in Carmel, Indiana (*13*). The result indicated that the calibrated model could accurately predict the capacity when the circulating flow rate was low or median but overestimated the capacity when the circulating flow rate was high. On an earlier study, they directly compared actual flow rates with the capacity model in the NCHRP Report 572, and found that it significantly underestimated the capacity (*14*).

Ren et al. calibrated five roundabout capacity models using the gap acceptance parameters measured at nine roundabouts in the Gold Coast, Australia (*15*). These calibrated models were then compared with the actual flow rates in the field. The calibrated HCM 2010 model underestimated the capacity by 4.62–16.14%.

The gap acceptance parameters on the same roundabout are not constant because driver behavior can be affected by several factors, such as the familiarity of the driver to the roundabout. Mensah et al. used a before-and after study to test the change in driver behavior by comparing critical headways in 2005 and in 2009 (*16*). The data were collected in two single-lane roundabouts in Maryland. The result showed that the critical headway reduced from 3.88 s in 2005 to 2.55 s in 2009. The study concluded that the critical headway reduced as the drivers became more experienced with the roundabout.

As the latest version of the highway capacity manual, the roundabout capacity model in HCM6 has not been evaluated in any study so far. The study presented in this paper evaluated the accuracy of the roundabout capacity model in HCM6 using the flow rate and gap acceptance data from a multilane roundabout in Richfield, MN. Two capacity curves were built to represent the relationship between the circulating flow and the entering flow, one for each lane. At first, the default roundabout capacity model in HCM6 was compared followed by a comparison between the field based curves and two calibrated versions of HCM6. One calibrated model had both parameters calibrated using observed critical and follow-up

Figure 1. Measurement locations.

headways, the other only calibrated the HCM6 relationship with the observed follow-up headway.

Data Collection and Processing

The data collection was based on video records. Videos were recorded by a 360-degree camera mounted on a mast at the center of the roundabout. Drivers' behaviors on all four approaches were captured simultaneously. The study is based on a total of 5 hours of observations collected, on each of the four roundabout approaches, during afternoon peak periods on October 10, 2016 (1 hour), March 16, 2017 (1.5 hours), March 17, 2017 (1 hour), and March 20, 2017 (1.5 hours), for a total of 20 hours of traffic observations.

The first set of information collected involved the times of four events, the entry of a vehicle to the roundabout on the right lane (E_R), the left lane (E_R), the passage of a circulating vehicle through the conflict point on the right lane (C_R), and the passage of a circulating vehicle through the conflict point on the left lane (C_L). These events are presented for one approach in Figure 1. During these events the vehicle type was also recorded which allowed for a later translation of flows to PCs/hour.

The second set of extracted information involved the time of queue initiation and suspension. A queue period is defined as the time interval between the first occurrence of an entering vehicle yielding to a circulating one and the time the last queued vehicle entered the roundabout.

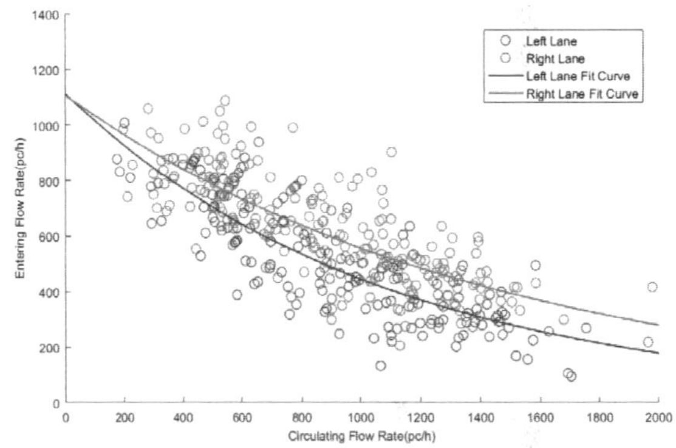

Figure 2. Field flow data and fitted capacity curves.

Flow Rate Data Extraction

Traffic volumes at capacity conditions were extracted on both lanes of the roundabout. For the purposes of this study, "at capacity" is defined as the period during which the queue on the left or right entry lanes has a duration of more than 30 s. Volumes were afterwards converted to hourly flow rates in pc/hour. In total, 159 data points for the left entry lane and 227 data points for the right entry lane were extracted. Figure 2 shows all of the collected data points. The blue points represent vehicles on the left entry lane and the red points represent vehicles on the right lane.

Gap Acceptance Data Extraction

From the available video records, critical and follow-up headway observations were extracted to use in the calibration of the HCM6 capacity model. Critical headway is the minimum headway between circulating vehicles an entering driver can safely use to enter the roundabout. Follow-up headway is the average time between consecutive queued vehicles entering the roundabout before being interrupted by a circulating one.

Estimation of Critical Headways. The Raff method (*17*) and the maximum likelihood method are the two most commonly used procedures for estimating critical headway. Both methods require accepted and rejected headways. A rejected headway is any headway between circulating vehicles not utilized by an entering vehicle waiting at the approach. These headways are based on driving behavior and can vary among drivers and locations.

In Raff's method, the critical headway is the headway that is equally likely to be accepted or rejected by drivers. To apply this method, two cumulative frequency distributions are developed, one for rejected headways and another for accepted headways. The intersection of the two cumulative distributions is the value of the critical headway (*17*).

The maximum likelihood method was used in the analysis described in the NCHRP Report 572. This method assumes that the size of the critical headway of a particular driver is between his/her accepted headway and his/her largest rejected headway. The distribution of the critical headway is assumed to be log-normal (*18*). To employ this method, observations of vehicles entering the roundabout are used if they meet three requirements: (1) the vehicle should be in a queue; (2) the vehicle should have rejected at least one headway before it accepts one and enters the roundabout; and (3) the largest rejected headway should be smaller than the accepted one (*18*).

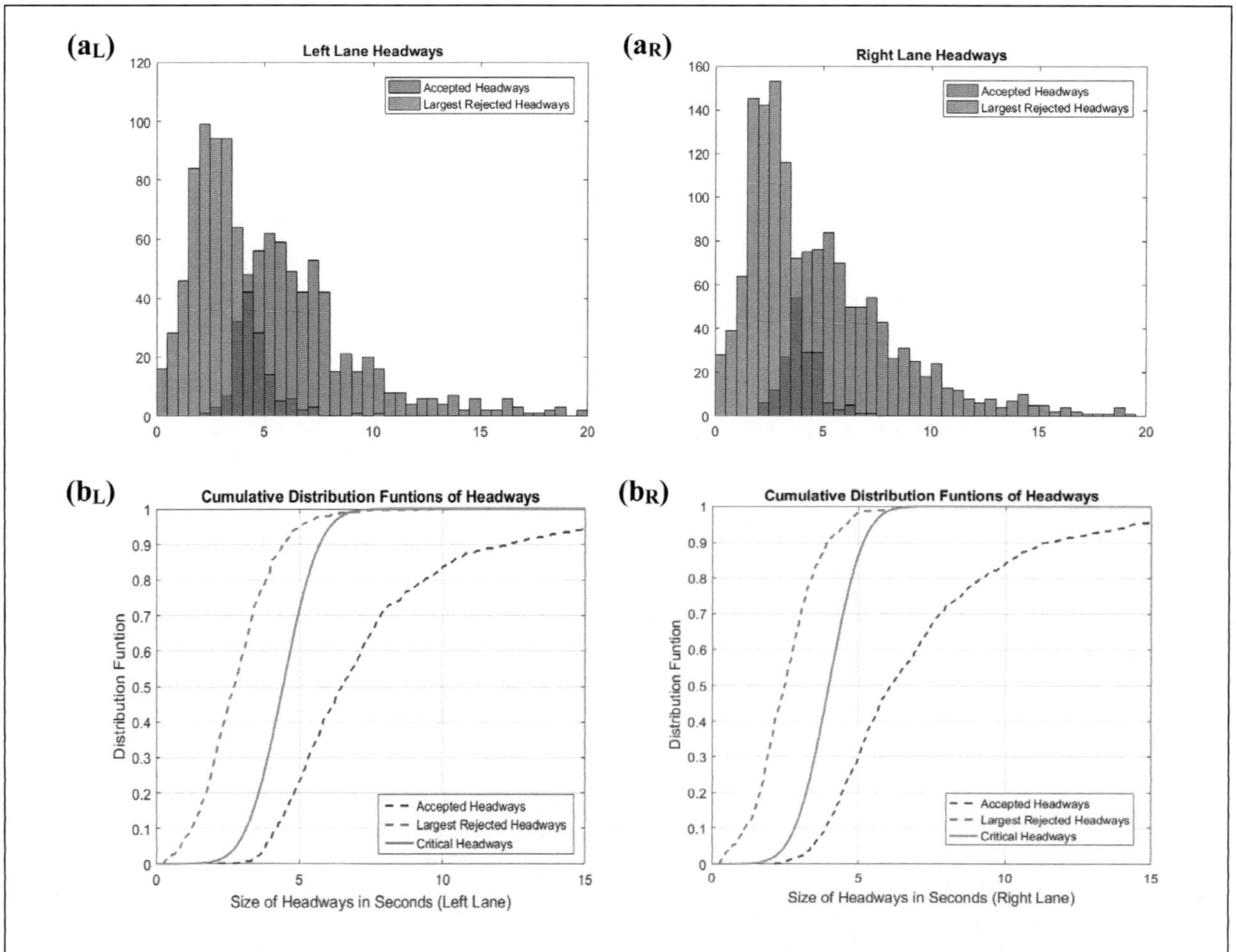

Figure 3. Observations of accepted and largest rejected headways. Histograms of collected data (a_L and a_R) and fitted cumulative distributions (b_L and b_R).

After filtering out unqualified observations in the dataset, there were 627 and 833 observations on the left and right lanes respectively. Figure 3, a_L and a_R show the histograms of the largest rejected and corresponding accepted headways for the left and right lanes respectively. Blue bars represent accepted headways and orange bars represent largest rejected headways. The most frequent largest rejected headways are between 2 and 3 s while the most frequent accepted headway is between 4 and 6 s.

In the maximum likelihood method, if the i_{th} driver has an accepted headway e^{y_i} and a largest rejected headway e^{x_i}, then a dataset with n drivers has a maximum log-likelihood L (*19*) given by:

$$L = \sum_{i=1}^{n} \ln(F(y_i) - F(x_i)) \quad (1)$$

where:

y_i = the logarithm of the gap accepted by the i_{th} driver

x_i = the logarithm of the largest gap rejected by the i_{th} driver

$F()$ = the cumulative distribution function of the normal distribution

The cumulative distribution function of a normal distribution is described by its mean μ and variance σ^2. The parameters of the normal distribution that describes critical headway are the ones that maximize L. The mean t_c and variance s^2 of the critical headway can be calculated from:

$$t_c = e^{\mu + 0.5\sigma^2} \quad (2)$$

$$s^2 = t_c\left(e^{\sigma^2} - 1\right) \quad (3)$$

For the samples collected in this study, estimates for μ and σ^2 were computed using the Python Package "scipy". The critical headway on the left lane had a mean of 4.428 s and a standard deviation of 0.985. The critical headway estimation on the right lane resulted in a mean of 3.992 s and a standard deviation of 0.905. Raff's method, not described here in detail, was also applied and resulted in estimates of 4.30 and 3.90 s for the left and right lanes respectively.

The following formula was used to test whether the sample size available was large enough to reach an accurate estimation:

$$n = \left[\frac{Z\alpha_{/2}\sigma}{E}\right] \quad (4)$$

where:

n = sample size

σ = population standard deviation

E = Margin of the error

$Z\alpha_{/2}$ = Z value for a given confidence interval α

For a confidence interval of 95%, and a 0.1 margin of error, the left and the right lanes required sample sizes of 373 and 315 observations respectively, which are smaller than the available observations used in this study; 627 for the left lane and 833 for the right lane.

Figure 3, b_L and b_R presents the cumulative distribution functions of the largest rejected headway, the accepted headway, and the estimated critical headway for the left and right lanes respectively. For the left lane, 95% of the largest rejected headways were smaller than 5 s and only 22% of the accepted headways were smaller than 5 s. 70% of the drivers had a critical headway smaller than 5 s. For the right lane, 97% of the largest rejected headway were smaller than 5 s while 30% of accepted headways were smaller than 5 s. 87% of the drivers had a critical headway smaller than 5 s.

Estimation of Follow-Up Headways. The estimation of the follow-up headway is a much simpler process as it is the mean of the collected observations. The frequencies of the observed follow-up headways are presented in Figure 4. There were 802 and 1127 collected follow-up headways on the left and right lanes respectively. The required sample sizes for a 95% confidence were 508 and 406, so the dataset was large enough to provide accurate estimates. The estimate of the follow-up headway on the left lane was 3.05 s while the same for the right lane was 2.96 s. Drivers on the left lane had larger follow-up headways than drivers on the right lane.

HCM6 and HCM 2010 Capacity Model Evaluation

Both linear and exponential regression model forms were tested for goodness-of-fit to the collected capacity flows. Table 1 lists the parameter estimation and goodness-of-fit scores. The exponential model provided a better fitting to the flow rate data than the linear model as suggested by the larger R^2 and smaller root-mean-square error (RMSE) reached.

The formulas for the exponential models, referred to in the rest of the paper as the Richfield models, for the left and right lanes are as follows:

$$C_{e,R,pce} = 1108e^{(-0.687 \times 10^{-3})v_{c,pce}} \quad (5)$$

$$C_{e,L,pce} = 1114e^{(-0.915 \times 10^{-3})v_{c,pce}} \quad (6)$$

Two measurements are used to calculate the difference between each of the HCM and the Richfield models. The formulas used are as follows:

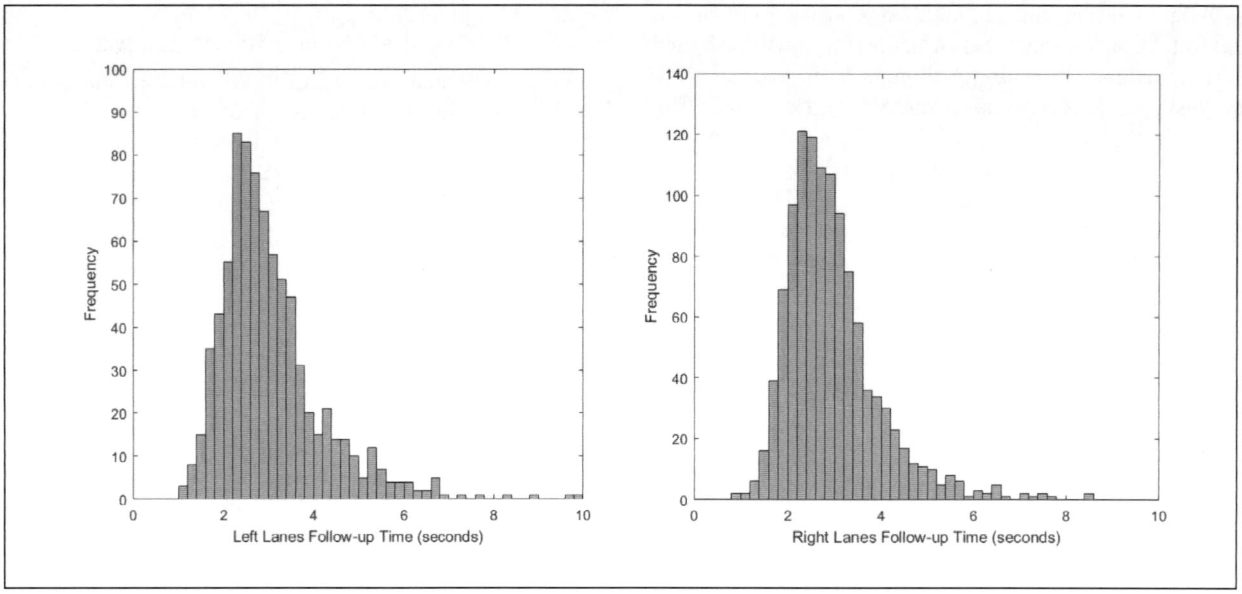

Figure 4. Frequency distribution of follow-up headways.

Table 1. Fitted Regression Models of Flows at Capacity

	p_1	p_2	SSE	R^2	RMSE
Linear model (left lane)	925.4	−0.4540	2.226e + 06	0.674	119.08
Linear model (right lane)	989.3	−0.4157	3.268e + 06	0.615	120.52
Exponential model (left lane)	1114	−0.0009151	2.151e + 06	0.685	117.10
Exponential model (right lane)	1108	−0.0006874	3.269e + 06	0.615	120.50

Note: SSE = sum of squared errors; RMSE = root-mean-square error.

$$\text{Capacity Difference} = \text{Model 1} - \text{Model 2} \quad (7)$$

$$\text{Relative Difference} = (\text{Model 1} - \text{Model 2})/\text{Model 2} \quad (8)$$

Default HCM6 Model versus Richfield Model

The comparison between the default two-lane roundabout capacity model in HCM6 and the model are presented in Figure 5. In Figure *5a*, the capacities estimated by the Richfield model are, in general, lower than the capacities produced by HCM6. For the right lane specifically, under the same circulating flow rate, the roundabout capacity in HCM6 is higher than the one estimated by the field data. When the circulating flow rate is higher than 1600 pc/h, the difference between the two models reduces. For the left lane, the capacity in HCM6 is always higher.

The blue and red shadow areas in Figure *5a* show the 95% confidence interval of the Richfield model for the left and the right lanes. As most parts of the default HCM6 models are outside the shadow areas, the default HCM6 model does not adequately represent the local conditions. The blue and red shadow areas in Figure *5b* show the 95% predictive confidence interval of the Richfield model. It shows that if the Richfield model is used to predict capacity given a value of circulating flow rate, 95% of the actual entering flow rates will be included in the predictive confidence interval. Figure *5c* shows the differences in capacity estimation under different circulating flow rates whereas Figure *5d* shows the relative difference of the model in HCM6 as compared with the Richfield one.

In particular, on the right lane, the difference in estimated capacity has a high value of about 310 pc/h when the circulating flow rate approaches zero. The relative difference starts at 30%, which means that the HCM6 model overestimates the capacity of the roundabout in Richfield when the circulating flow rate is very low. As the circulating flow rate increases, the capacity difference between the two models, as well as the relative differences decrease and approach zero when the circulating flow is around 1600 pc/h. On the left lane, the difference between the capacity models starts at 240 pc/h and drops

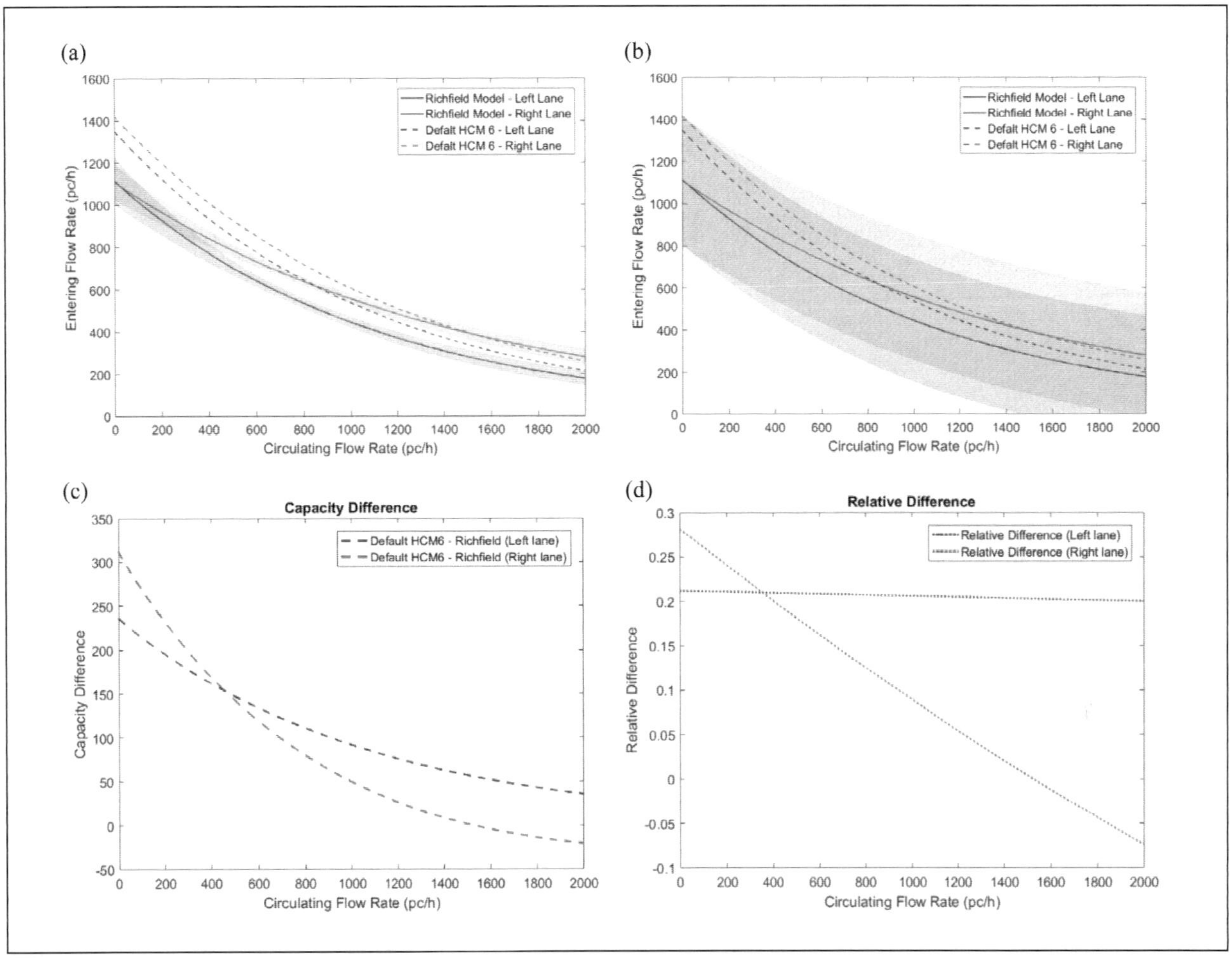

Figure 5. Comparison between the HCM6 and Richfield models: (a) comparison of entering flows between default HCM6 and Richfield model with 95% confidence interval of Richfield model; (b) comparison of entering flows between default HCM6 and Richfield mode with 95% predictive confidence interval of Richfield model; (c) the absolute difference between entering flows of default HCM6 and Richfield model; (d) the relative difference between entering flows of default HCM6 and Richfield model.

to 50 pc/h as the circulating flow rate reaches 2000 pc/h. According to the graph of the relative difference, the capacity model in HCM6 always overestimates the capacity of the roundabout in Richfield by approximately 20% regardless of the level of circulating flow.

Default HCM 2010 Model versus Richfield Model

The comparison between the default two-lane roundabout capacity models in HCM 2010 and the model fit on the field collected data are presented in Figure 6. In Figure 6, *a* and *b* the capacities estimated by the Richfield model are almost identical for the right lane. For the left lane, HCM 2010 overestimates the capacity of the Richfield roundabout by 2–41% as the circulating flow increases from 0 to 2000 pc/h.

HCM Model Calibration

Based on the roundabout capacity formula in the HCM, given locally collected values of the critical and follow-up headways, the two parameters of the exponential model can be estimated and produce a locally calibrated capacity model. There are two options in calibrating the HCM model: calibrate both parameters A and B using the critical and follow-up headways, or calibrate only parameter A using the follow-up headway. The HCM model form is as follows:

$$C_{pce} = Ae^{(-Bv_c)} \tag{9}$$

$$A = \frac{3600}{t_f} \tag{10}$$

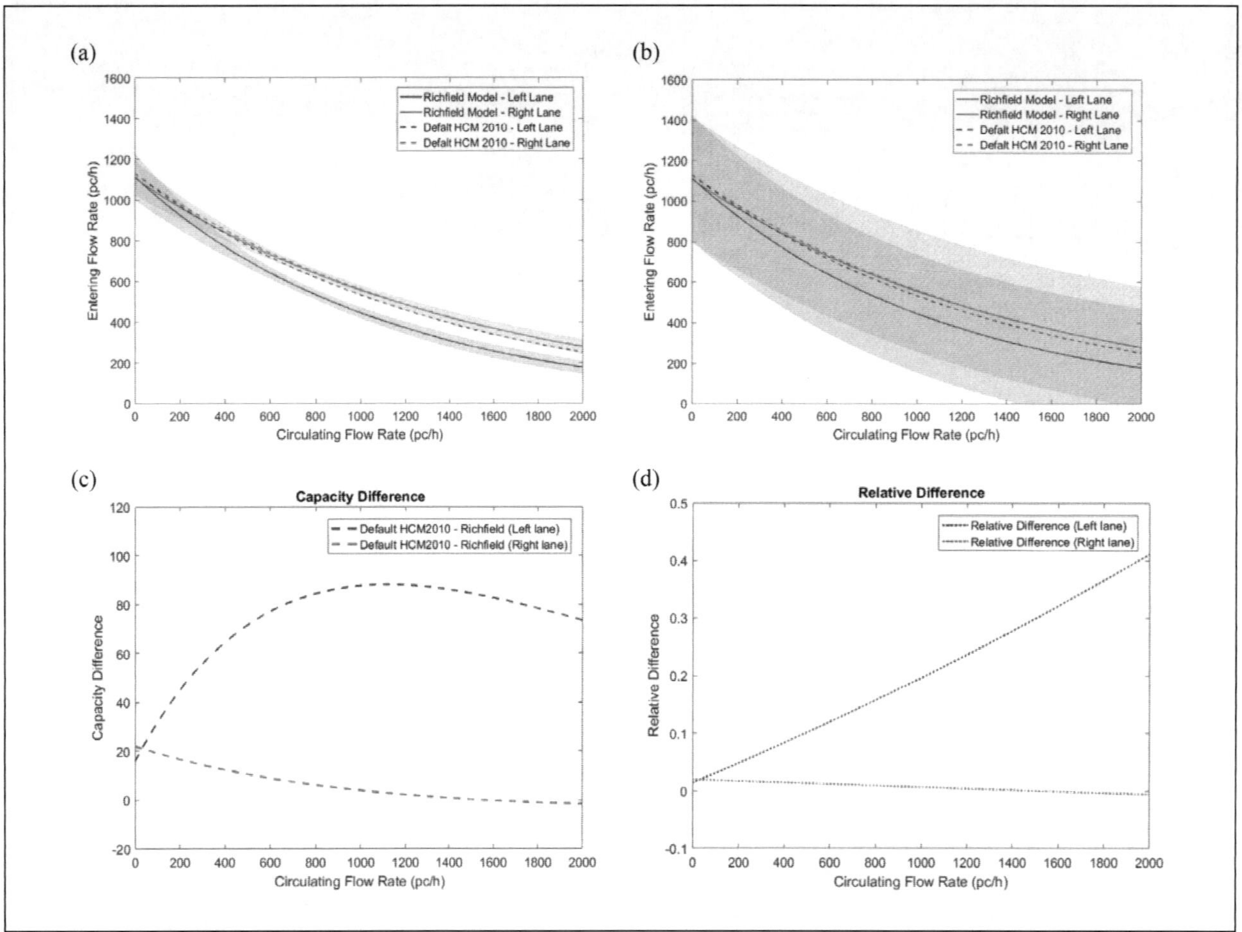

Figure 6. Comparison between the HCM 2010 and Richfield models: (a) comparison of entering flows between default HCM 2010 and Richfield model with 95% confidence interval of Richfield model; (b) comparison of entering flows between default HCM 2010 and Richfield mode with 95% predictive confidence interval of Richfield model; (c) the absolute difference between entering flows of default HCM 2010 and Richfield model; (d) the relative difference between entering flows of default HCM 2010 and Richfield model.

Table 2. Comparison of Observed and Modeled Headways

		Critical headway	Follow-up headway
Richfield model	Left lane	4.910	3.232
	Right lane	4.099	3.249
Observed	Left lane	4.428	3.049
	Right lane	3.992	2.964
HCM6	Left lane	4.650	2.667
	Right lane	4.320	2.536
HCM 2010	Left lane	4.293	3.186
	Right lane	4.113	3.186

$$B = \frac{t_c - (t_f/2)}{3600} \tag{11}$$

where:

C_{pce} = lane capacity
v_c = conflicting flow (pc/h)
t_c = critical headway (s)
t_f = follow-up headway (s)

Parameter A is the intercept, which only relates to the follow-up headway. A shorter follow-up headway results in a larger intercept. Table 2 presents the different headways resulting from the different models, as well as the ones observed in the field. The follow-up headway in HCM6 is the shortest, resulting in the highest capacity as the circulating flow approaches zero. Parameter B is the slope which controls the rate of change. A longer critical headway and a shorter follow-up headway can lead to a rapidly decreasing capacity. Among the three models, the HCM6 has the greatest rate of decrease.

Figure 7 compares the values produced by the default HCM6 model with the fully calibrated HCM (with both

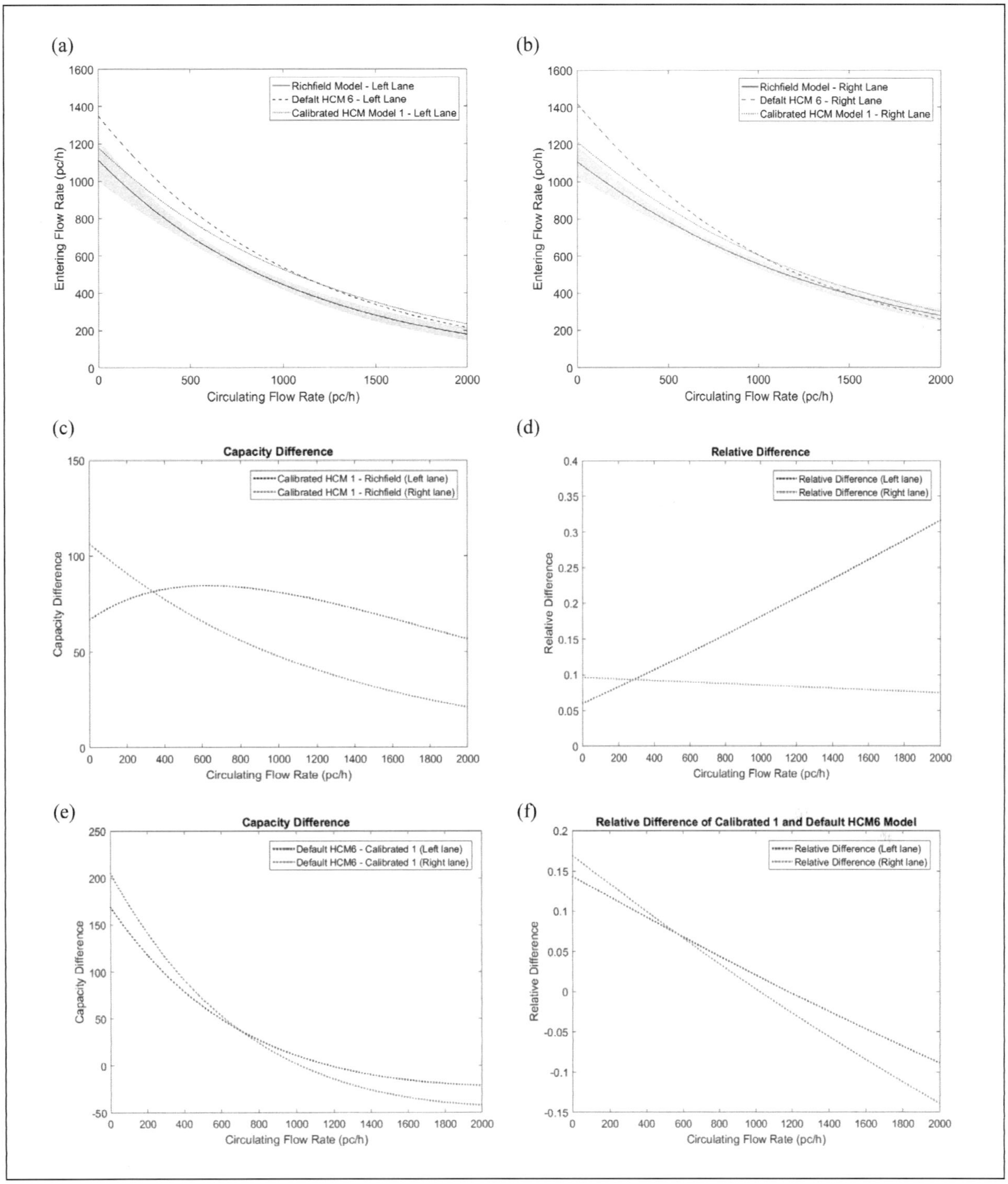

Figure 7. Comparison between the default HCM6 and the fully calibrated HCM models: (a) comparison of entering flows between default HCM6 model, fully calibrated HCM model, and Richfield model with 95% confidence interval of Richfield model (left lane); (b) comparison of entering flows between default HCM6 model, fully calibrated HCM model, and Richfield model with 95% confidence interval of Richfield model (right lane); (c) the absolute difference between entering flows of fully calibrated HCM model and Richfield model; (d) the relative difference between entering flows of fully calibrated HCM model and Richfield model; (e) the absolute difference between entering flows of default HCM6 model and fully calibrated HCM model; (f) the relative difference between entering flows of default HCM6 model and fully calibrated HCM model.

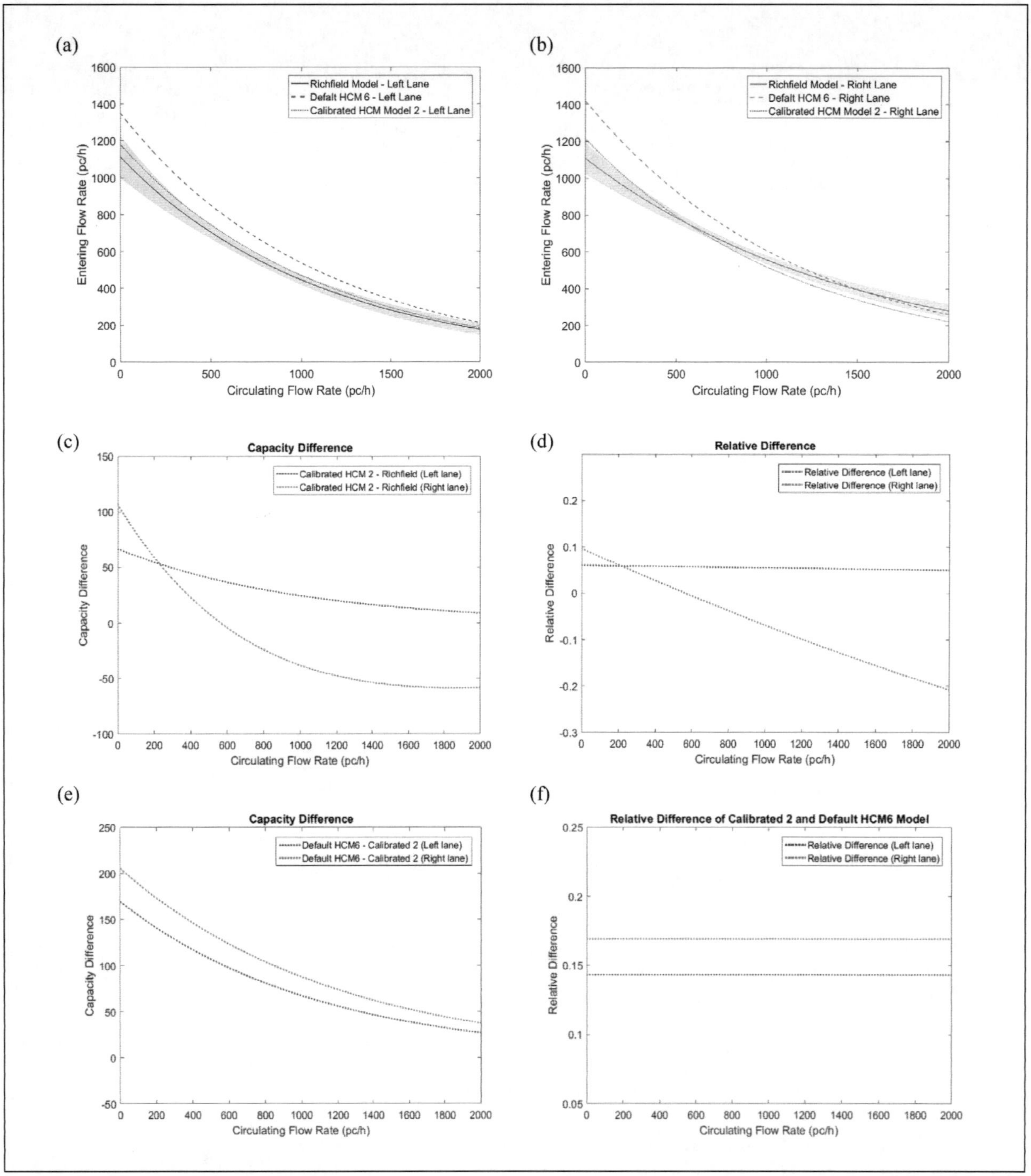

Figure 8. Comparison between the default HCM6 and the partially calibrated HCM models: (a) comparison of entering flows between default HCM6 model, partially calibrated HCM model, and Richfield model with 95% confidence interval of Richfield model (left lane); (b) comparison of entering flows between default HCM6 model, partially calibrated HCM model, and Richfield model with 95% confidence interval of Richfield model (right lane); (c) the absolute difference between entering flows of partially calibrated HCM model and Richfield model; (d) the relative difference between entering flows of partially calibrated HCM model and Richfield model; (e) the absolute difference between entering flows of default HCM 6 model and partially calibrated HCM model; (f) the relative difference between entering flows of default HCM6 model and partially calibrated HCM model.

parameters recalculated). When the circulating flow rate is low, the fully calibrated model experiences less difference from the Richfield model. The capacity difference between the calibrated HCM model and the Richfield model under low circulating flow rate (<300 pc/h) is between 80 to 110 pc/h, with a relative difference no larger than 10%. In contrast, under heavy circulating flow, the roles are reversed with the default model experiencing less difference from the Richfield model than the calibrated one. Under moderate-to-high circulating flow rates (>300 pc/h), the capacity difference between the calibrated model and the Richfield model is between 25 to 80 pc/h.

For the left lane, the capacity difference between the calibrated model and the Richfield model is always between 50 to 80 pc/h while the capacity difference between the default HCM6 model and the Richfield model is between 40 to 240 p/h. Therefore, the calibrated model for the left lane is also closer to the Richfield model than the default HCM6 model. Therefore, the fully calibrated models for both lanes are closer to the Richfield model than the default HCM6 model.

Calibrating parameter A only requires the collection of follow-up headways and therefore involves much less effort. Figure 8 compares the values produced by the default HCM6 model with the partially calibrated model. Surprisingly, the partially calibrated model produced results that are closer to the exponential regression model produced from the observed flows.

On the right lane, the capacity difference starts at 110 pc/h and reduced to −60 pc/h as the circulating flow rate increases. The relative difference ranges between 10% and −21%. On the left lane, the capacity difference of the partially calibrated HCM (model 2) starts at 60 pc/h and ends at 10 pc/h with a relative difference remaining steady around 6%.

Overall, the calibration of the HCM model is worthwhile because both calibrated HCM models are more accurate than the default. Given the results presented, performing only partial calibration by collecting follow-up headways is sufficient and possibly better than a calibration based on both critical and follow-up headways.

Conclusion

This study extracted the traffic flow rates and gap acceptance data from a multilane roundabout in Richfield, Minnesota. Two models were produced based on the collected flow data, one for each entrance lane. These local models were compared with the default models proposed in HCM 2010 and the latest HCM6. Specifically, for the HCM6 model, the right lane capacity was overestimated by up to 28%. For the left turn, the HCM6 again overestimated the capacity, but no more than 20%.

Based on the collected gap acceptance information, the HCM model was calibrated by using both the critical and follow-up headways (parameters *A* and *B*), as well as by calibrating the intercept (parameter A) based on the collected follow-up headways. The partially calibrated model experienced smaller differences from the field data based model as compared with the differences between the field data based model and the fully calibrated HCM model. Therefore, the model calibration is useful to improve the accuracy of the default model and calibrating only the intercept of the model can produce a model with higher accuracy.

Author Contributions

The authors confirm contribution to the paper as follows: study conception and design: Rongsheng Chen; data collection: John Hourdos, Rongsheng Chen; analysis and interpretation of results: Rongsheng Chen; draft manuscript preparation: Rongsheng Chen, John Hourdos. All authors reviewed the results and approved the final version of the manuscript.

References

1. *SIDRA Intersection User Guide*. Akcelik and Associates Pty Ltd., Victoria, Australia, 2010.
2. *RODEL 1 Interactive Roundabout Design*. RODEL Software Ltd and Staffordshire County Council, UK, 2002.
3. *TORUS Roundabouts 4.0*. Transoft Solutions Inc., Richmond, Canada, 2012.
4. *Highway Capacity Manual: A Guide for Multimodal Mobility Analysis*, 6th ed. Transportation Research Board, Washington, D.C., 2016.
5. *Highway Capacity Manual 2010*. Transportation Research Board of the National Academies, Washington, D.C., 2010.
6. Hourdos, J., and V. Richfield. *Effect of Signing and Lane Markings on the Safety of a Two-Lane Roundabout*. Report to MN Department of Transportation, University of Minnesota, Twin Cities, MN, 2014.
7. Rodegerdts, L., M. Blogg, E. Wemple, E. Myers, M. Kyte, M. P. Dixon, G. F. List, A. Flannery, R. Troutbeck, W. Brilon, N. Wu, B. N. Persaud, C. Lyon, D. L. Harkey, and D. Carter. *NCHRP Report 572: Roundabouts in the United States*. Transportation Research Board of the National Academies, Washington, D.C., 2007.
8. Rodegerdts, L., A. Malinge, P. Marnell, S. Beaird, M. Kittelson, and Y. Mereszczak. *Accelerating Roundabouts in the United States: Volume II of VII - Assessment of Roundabout Capacity Models for the Highway Capacity Manual*. U.S. Department of Transportation, Federal Highway Administration, Washington, D.C., 2015.
9. Xu, F., and Z. Tian. Driver Behavior and Gap-Acceptance Characteristics at Roundabouts in California. *Transportation Research Record: Journal of the Transportation Research Board*, 2008. 2071: 117–124.

10. Zheng, D., M. V. Chitturi, A. R. Bill, and D. A. Noyce. Critical Gaps and Follow-up Headways at Congested Roundabouts. Presented at 91st Annual Meeting of the Transportation Research Board, Washington, D.C., 2012.
11. Gazzarri, A., M. Martello, A. Pratelli, and R. Souleyrette. Gap Acceptance Parameters for HCM 2010 Roundabout Capacity Model Applications in Italy. *Wit Serifes on Transport Systems & Traffic Engineering*, Vol. 1, 2013, pp. 1–6.
12. Fitzpatrick, C., D. Abrams, Y. Tang, and M. Knodler. Spatial and Temporal Analysis of Driver Gap Acceptance Behavior at Modern Roundabouts. *Transportation Research Record: Journal of the Transportation Research Board*, 2013. 2388: 14–20.
13. Wei, T., and J. Grenard. Calibration and Validation of Highway Capacity Manual 2010 Capacity Model for Single-Lane Roundabouts. *Transportation Research Record: Journal of the Transportation Research Board*, 2012. 2286: 105–110.
14. Wei, T., J. Grenard, and H. Shah. Developing Capacity Models for Local Roundabouts: Streamlined Process. *Transportation Research Record: Journal of the Transportation Research Board*, 2011. 2257: 1–9.
15. Ren, L., X. Qu, H. Guan, S. Easa, and E. Oh. Evaluation of Roundabout Capacity Models: An Empirical Case Study. *Journal of Transportation Engineering*, Vol. 142, No. 12, 2016, p. 04016066.
16. Mensah, S., S. Eshragh, and A. Faghri. A Critical Gap Analysis for Modern Roundabouts. Presented at 89th Annual Meeting of the Transportation Research Board, Washington, D.C., 2010.
17. Raff, M. S., and J. W. Hart. A Volume Warrant for Urban Stop Sign. *Traffic Engineering and Control*, Vol. 5/1983, 1950, pp. 255–258.
18. Troutbeck, R. J. *Estimating the Critical Acceptance Gap from Traffic Movements*. Research Report 92–5. Physical Infrastructure Centre, Queensland University of Technology, Brisbane, Australia, 1992.
19. Tian, Z., M. Vandehey, B. W. Robinson, W. Kittelson, M. Kyte, R. Troutbeck, W. Brilon, and N. Wu. Implementing the Maximum Likelihood Methodology to Measure a Driver's Critical Gap. *Transportation Research Part A: Policy and Practice*, Vol. 33, No. 3, 1999, pp. 187–197.

The Standing Committee on Highway Capacity and Quality of Service (AHB40) peer-reviewed this paper (18-01297).

Article

TRR
JOURNAL OF THE TRANSPORTATION RESEARCH BOARD

Transportation Research Record
2018, Vol. 2672(15) 35–43

DOI: 10.1177/0361198118777603
journals.sagepub.com/home/trr
SAGE

Investigating Entry Capacity Models of Roundabouts under Heterogeneous Traffic Conditions

Ashish Kumar Patnaik[1], Ankit Raj Ranjan[1], and Prasanta Kumar Bhuyan[1]

Abstract

The primary objectives of this study are to develop two roundabout entry capacity models using a regression-based multiple non-linear regression model (MNLR) and artificial intelligence (AI)-based ANFIS (adaptive neuro-fuzzy inference system) model under heterogeneous traffic conditions. ANFIS is the latest technique in the field of AI that integrates both neural networks and fuzzy logic principles in a single framework. Required data have been collected from 27 roundabouts in eight states of India. To assess the significance of these models and select the best model among them, modified rank index is applied in this study. The coefficient of determination (R^2) and Nash–Sutcliffe model efficiency coefficient 'E' values are found to be 0.92, 0.91 and 0.98, 0.98 for the MNLR and ANFIS model, respectively. The ANFIS model is found to be the best model in this study. However, from a practical point of view, the MNLR model is recommended for determining roundabout entry capacity under heterogeneous traffic conditions. Sensitivity analysis reports that critical gap is the prime variable and shares 18.43% for the development of roundabout entry capacity. As compared with the Girabase formula (France), Brilon wu formula (Germany), and HCM 2010 models, the proposed MNLR model is quite reliable under low to medium ranges of traffic volumes.

Roundabouts are a specialized form of an un-signalized intersection in which traffic moves in one direction around a central island. Direction of traffic, according to left-hand or right-hand driving rules, depends on the specific country. Roundabouts have been used as an effective intersection control measure to improve safety and operational performance under low to high circulating traffic flow conditions. As compared with other, signalized, intersections, roundabouts have several basic advantages, such as speed reduction of vehicles, less delay, pedestrian safety, aesthetics, and reduction of conflict points. A signalized intersection has 32 conflict points, whereas a roundabout with two circulating and entry lanes has 8 and 16 conflict points, respectively. Because of the increase in the use of roundabouts, an assessment regarding operational performance needs to be undertaken. Evaluation of capacity at roundabouts is an important aspect of operational performance that describes the present traffic scenario, level of service, delay, accidents, operational cost, and environmental issues.

Under heterogeneous traffic conditions, vehicles of various kinds have varying physical and operational characteristics. In developing countries, driver behavior, vehicular characteristics (size, engine efficiency, and years of usage), and intersection geometric design vary significantly. The vehicle fleet is composed of heavy vehicles (HV), light motor vehicles (LMV), motor cycles/scooters (MC/S), bicycles (BC) and human-drawn vehicles (HDV). These vehicles share the main carriageway without any physical or marked segregation. When there are vehicles of different types with varying static and dynamic characteristics, follow-the-lane discipline and the follow-the-leader concept become irrelevant. Under constraints of flow and space, small-sized vehicles try to move into the circulating area without following lane discipline. Also, HV forcibly merge into the circulating stream of traffic. These points reflect that the rule of priority is not respected under such a heterogeneous traffic condition. In these circumstances, the study of

[1]Department of Civil Engineering, NIT Rourkela, India

Corresponding Author:
Address correspondence to Ashish Kumar Patnaik: ashishkumar.patnaik@yahoo.in

gap-acceptance behavior is highly complex in nature. Since gap-acceptance variables are treated as major variables for the determination of roundabout entry capacity, critical gap values should be carefully observed under heterogeneous traffic conditions. Inaccurate estimation of critical gap values leads to biased capacity values. Also under heterogeneous traffic conditions, the critical gap values are marginally different as compared with homogeneous traffic flow conditions. This is attributed to the difference in vehicle maneuverability and the aggressiveness of Indian drivers. Thus the homogeneous traffic flow condition omits some important aspects of heterogeneous traffic conditions.

Background Literature

The capacity of each entry to a roundabout is the maximum flow rate at which vehicles can reasonably be expected to enter the roundabout during a given time period under prevailing traffic and geometric conditions. Existing capacity methods are broadly categorized into two groups, empirical and gap-acceptance models. In empirical regression-based models, capacity is taken as a dependent variable, whereas the circulating flow and other geometric variables are taken as independent variables (*1*, *2*). The gap-acceptance capacity model considers the driver's decision to move from the minor stream of traffic flow to the major stream of traffic flow. Critical gap and follow-up gap are the two important variables for the development of the gap-acceptance capacity model. HCM 2000 is a popular reference for roundabout capacity that follows the concept of gap acceptance (*3*). HCM 2010 gives a detailed procedure for determining the capacity and level of service analysis of roundabouts (*4*). The model is a combination of gap acceptance and the exponential regression concept, and is calibrated by estimating critical gap and follow-up gap.

In order to develop the roundabout entry capacity models, Rodegerdts et al. found that prior knowledge regarding the relationship between dependent and independent variables is required, and found difficulties in investigating the relationship owing to scatter type of data (*5*). In the recent past, researchers found that artificial intelligence (AI)-based methods were able to forecast the data more accurately as compared with regression-based models. Gupta et al. measured the performance of plate-fin heat exchangers using different AI techniques, and found that the adaptive neuro-fuzzy inference system (ANFIS) model predicts the data more accurately as compared with other artificial neural network (ANN), genetic algorithm and simulated annealing models (*6*). Therefore, the literature provides a strong background regarding roundabout capacity formulation in which AI-based modeling predicts the capacity value more accurately in comparison with gap-acceptance and empirical-based models under a new data set.

Study Objectives and Organization of the Paper

With an overview of heterogeneous traffic conditions in developing countries, the following study objectives are provided below.

- The objectives of the study are to develop two roundabout entry capacity models, one a regression-based model and other based on an AI technique such as the ANFIS model, to check model transferability and prediction ability.
- In order to reflect the heterogeneity of traffic flow, different compositions of vehicles are converted to single car unit (PCU). Also the critical gap and follow-up time is estimated by an influence area for gap acceptance (INAGA) method for accurate estimation of capacity values under heterogeneous traffic conditions.
- To assess the significance of the two models and select the best model under heterogeneous traffic conditions, modified rank index (MRI) is applied.
- To analyze the relative importance of input variables in the model development, sensitivity analysis is carried out.

According to these objectives, the paper is categorized as follows. The section "Methodology" describes critical gap and follow-up time estimation, modeling methods to determine roundabout entry capacity, and details regarding study area and the data collection procedure. Development of roundabout entry capacity models based on regression analysis and AI-based technique ANFIS, significance test of developed models, sensitivity analysis, and comparison of models are described in the "Results and discussion" section. Finally, the practical application of this research, limitations of the study, and direction for further research are discussed in the "Conclusion."

Methodology

The study is categorized into two parts. The first part includes the determination of gap-acceptance parameters (critical gap and follow-up time) by employing an INAGA method. In the second part two roundabout entry capacity models are developed using regression analysis and AI technique ANFIS.

Determination of Critical Gap and Follow-Up Time

The driver behavior variables critical gap and follow-up time are the two prime variables for the development of roundabout entry capacity models. Critical gap measurement is a complex phenomenon which is based on the assessment of driver behavior. Several studies have been carried out for the determination of critical gap. Raff and Hart defined the critical gap as the size of the gap for which the number of accepted gaps shorter than it was equal to the number of rejected gaps longer than it (*7*). HCM 2000 reported that the critical gap is the minimum gap in the circulating path in which the vehicle can safely enter the roundabout (*3*). Follow-up time is taken as the average time gap between two consecutive vehicles at the entry leg. Existing methods, such as the latest developed equilibrium probabilities method (*8*) and traditionally followed Raff method (*7*), assume that the driver is consistent and traffic flow is homogeneous for estimation of the critical gap. But these methods do not give reliable results under no rejection of gap data. This fact is therefore seen as a major issue under heterogeneous traffic conditions. Generally, no gaps are rejected for heavy and free-flow traffic under heterogeneous traffic conditions. Thus, to address this problem and heterogeneity, the INAGA method is used for the determination of critical gap and follow-up time in this study (*9*).

Adaptive Neuro-Fuzzy Inference System

The basic framework behind the ANFIS technique is simply a fuzzy inference system which is successfully implemented in the framework of adaptive networks, and so emerges a new technique of AI termed ANFIS (*10*). It is useful in determining the relation among dependent and independent variables by using a hybrid learning technique. In this study the backpropagation error method, that is, hybrid algorithm, is used to train the ANFIS-based model. The proposed model can easily construct an input–output relation-based mapping using human knowledge and provided or stipulated input–output data pairs.

Study Sites and Data Collection Procedure

In order to develop the two roundabout entry capacity models, 27 roundabouts were chosen for the purpose of data collection. These roundabouts are located in eight states of India, covering northern, southern, eastern and western zones, and details are given in Figure 1. Selected roundabouts are situated at grade intersection with no rise and fall (Rise/fall = 0). Roundabouts are uncontrolled in nature, having one to four lanes in entry and circulating path, respectively. The following criteria have been taken into consideration for selecting the roundabouts: the roundabouts are free from the effect of curvature, side friction, parking spaces, bus stops, and pedestrian movements.

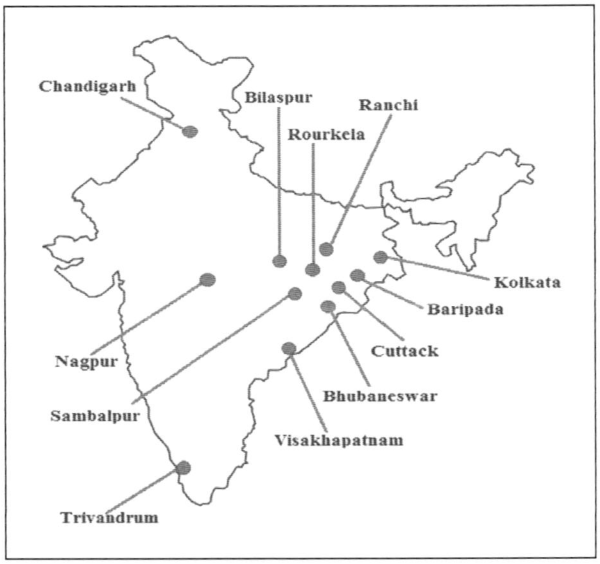

Figure 1. Location of study areas on Indian map.

Video shooting technique was used for data collection. Two high-resolution video cameras were fitted to a high-rise building with the help of a tripod stand to record the traffic flow variables and extract gap-acceptance variables such as critical gap and follow-up time to develop the roundabout entry capacity models. The observations were made for a duration of at least 2 hours, under which more than half an hour was under queued condition at approach legs. Data collection was carried out in the months of October–February 2014. These months are considered as normal months for which the traffic flow is least affected by environmental influences. The videos were captured either from 9 am to 11 am or from 5 pm to 7 pm on a typical clear weekday. Sixty hours of videos were recorded for this study, and these videos were played in the laboratory to extract the desired data. Geometric variables were measured using measuring tape and measuring wheel during the early morning to avoid obstruction of the traffic flow. To maintain the accuracy of geometric variables, the measurements were verified by Google Earth software. After acquiring an extensive and a legitimate data set, traffic flow, gap-acceptance variables, and road geometrics were employed in regression analysis and AI-based techniques such as the ANFIS model to develop the two roundabout entry capacity models, and details are given in Table 1. To reflect the heterogeneity of traffic flow conditions, vehicles are classified into five types: HV, LMV, MC/S, BC, and HDV. Then, the total numbers of entry flow and circulating flow were converted into static PCU as given on IRC: 65-1976 (*11*). PCU factors are generally found by considering the vehicle categories and their composition,

Table 1. Investigation of Explanatory Variables in this Study

Variables	Units	Minimum	Maximum	Mean	SD
Observed entry capacity (Q_e)	PCU/h	250	3415	1819.6	774.85
Circulating flow (q_c)	PCU/h	219	3688	1071.29	721.52
Weaving length (W_l)	M	28.96	58.62	44.49	6.15
Weaving width (W_w)	M	8.2	20.05	17.90	5.69
Entry width (E_w)	M	5	20.12	15.18	3.90
Diameter of central island (D)	M	10.76	60.12	43.63	11.78
Critical gap (T_c)	Seconds	0.54	2.87	1.83	0.59
Follow-up time (T_f)	Seconds	0.92	2.99	1.58	0.53

Note: SD = Standard deviation of the sample size.

maneuverability, and speed of the vehicles, which varies from one country to another.

HV such as trucks, buses, and light commercial vehicles run on Indian roads integrating the configuration such as 179 HP@2500 rpm in which the kerb weight is 5970 kg with an overall dimension that varies from 7 to 12.5 m. HDV considered for this study are three-wheeled pedal-cycle rickshaws that carry a maximum of two passengers and three-wheeled pedal-cycles that carry a small amount of goods over a short distance. These HDV share the main stream carriageway without any physical or marked segregation from motorized traffic. Average length and width of LMV varies from 4–4.5 m and 1.6–1.7 m, respectively. It was observed that the traffic flow varies from 240 (free-flow condition) to 5683 (capacity condition) at the selected roundabouts. The composition of the traffic varied as 0.26–61%, 10–79.5%, 14.55–79%, 0.05–33.1%, 0.1–7% for HV, LMV, MC/S, BC, and HDV, respectively, at the selected roundabouts. It was found that the percentage of traffic share of each category vehicles was not greater than 80%. Therefore, it is concluded from the analysis that the traffic flow is widely heterogeneous traffic in nature (*12*).

Results and Discussion

In this section, development of the roundabout entry capacity models using regression analysis and the AI-based technique ANFIS is described. To assess the prediction performance and select the best model among them, MRI is applied. Sensitivity analysis is carried out to examine the percentage contribution of each explanatory variable in the model development. Finally, the developed models are compared with existing international models to ascertain the prediction capability under heterogeneous traffic conditions.

Development of Roundabout Entry Capacity Models

Regression-Based Model. Using the available data, a new roundabout entry capacity model was developed in SPSS software by employing multiple non-linear regression

Table 2. Statistical Characteristics of MNLR Model of Roundabout

Analysis of variance (ANOVA)

Model	Sum of squares	Degree of freedom	Mean square	F	Significant value
Regression	23.694	7	3.385	218.141	0.000
Residual	1.071	69	0.016		
Total	24.765	76	3.401		

R^2= 0.94, Adjusted R^2= 0.952

Estimation of variables by MNLR analysis

Model variables	Variable's coefficient	Standardized error	t-statistic value	Significant value ($p < 0.05$)
Circulating flow (q_c)	−0.013	0.004	−2.926	.005
Critical gap (T_c)	0.318	0.076	4.191	.000
Follow-up gap (T_f)	−0.127	0.054	−2.334	.022
Weaving length (W_l)	0.781	0.184	4.250	.000
Weaving width (W_w)	0.019	0.006	3.309	.001
Entry width (E_w)	0.023	0.008	3.094	.003
Diameter of central island (D)	−0.086	0.038	−2.246	.028

Table 3. ANFIS Parameters Details

No. of fuzzy rules	No. of I/P membership functions	Membership functions I/P	Membership functions O/P	Defuzzification method	R^2
2187	3,3,3,3,3,3,3	Triangular	Constant	Weighted sum	0.968
5184	3,3,3,3,4,4,4	Trapezoidal	Linear	Weighted average	0.946
2592	3,3,3,4,4,2,3	Sigmoid	Constant	Weighted average	0.932
4096	4,4,4,4,4,2,2	Triangular	Constant	Weighted sum	0.956
3072	2,2,4,3,4,4,4	Gbell	Linear	Weighted average	0.983

Note: I/P = input; O/P = output.

(MNLR) analysis. The model describes the relationship between independent explanatory variables and dependent variable to minimize the errors (residuals). Before developing the model, a preliminary assessment was carried out to identify the dependency of individual variables with regard to the dependent variable. In this study, 110 data points (observations from each approach leg of roundabouts) were used. Out of 110, 70% of data (77 data points) was used for model development and the remaining 30% (33 data points) was used for model validation.

Kutner et al. (*13*) observed that a non-linear relationship could be more appropriate without prior knowledge of the relationship between dependent and independent variables. Therefore the MNLR model was obtained by taking weaving length (W_l), critical gap (T_c), weaving width (W_w), entry width (E_w), circulating flow (q_c), follow-up gap (T_f), and diameter of central island (D) as independent variables and entry capacity (Q_e) as the dependent variable, as shown in Equation 1.

$$Q_e = 55.92 \times (D)^{-.086} \times (W_l)^{.781} \times (T_c)^{.318} \times e^{(.019W_w + .023E_w)} \times e^{-(\frac{.013q_c}{100} + .127T_f)} \quad (1)$$

(R^2 = 0.94, Adjusted R^2 = 0.952, Standard error of estimate = 0.124).

The statistical characteristics of the MNLR model show that the model is statistically significant at the 95% confidence level by integrating all the variables together, shown in Table 2. The R^2 of the model is found to be 0.94, which signifies the better fitness of the model. The p-value in the analysis of variance (ANOVA) test is found to be 0.00 (p-value < 0.05) and the standard error of estimate is 0.124, which signify that the model is statistically fit. Different statistical tests were then conducted to determine the contribution of each explanatory variable in the model development. In this study, the standard errors of different variables were found to range between 0.004 and 0.184, which is basically close to zero. The significance of each variable was tested through significant value (p-value < 0.05) and t-statistic value, which are observed to be significant in the proposed model. At the 95% confidence level (p-value < 0.05), the t-statistic value should be greater than $\pm$ 1.960. In this study, the t-statistic value of all seven variables was found to be greater than $\pm$ 1.960, which indicates that these variables are statistically significant. Also the significance (p-value) values of independent variables weaving length (W_l), critical gap (T_c), weaving width (W_w), entry width (E_w), circulating flow (q_c), follow-up gap (T_f), and diameter of central island (D) were found to be 0.00, 0.00, 0.001, 0.003, 0.005, 0.022, and 0.028, respectively, which are below 0.05 (5%). Therefore it was determined that the variables are contributing to model development at more than the 95% confidence level.

Artificial Intelligence-Based Model. ANFIS is the latest model in the field of AI that uses the concepts of both ANN and fuzzy inference system (FIS). In this system, prior knowledge is required for the selection and number of membership function for each individual variable. In ANN modeling, the backpropagation technique is used for training of input data, while ANFIS uses the hybrid technique to train the data set. The hybrid technique uses least square error in addition to backpropagation error for training. The model utilizes the advantages of both ANN and FIS to adjust its variables and obtain the optimum solution. Users can easily ascertain the prediction ability of model and the complexity of the network by selecting the number of hidden layers and the number of nodes. Thus, to assess the prediction values of roundabout entry capacity, an AI-based ANFIS model was applied for comparative purposes in this study.

ANFIS Model. In this study, the ANFIS model was developed to obtain the non-linear association between input and output variables of roundabout entry capacity using the neuro-fuzzy toolbox integrated in MATLAB version R 2013 a. The above seven variables are given as an input to the ANFIS model in MATLAB (R 2013 a) to establish the relationship between these input variables and entry capacity (Q_e) of the roundabout. There are five different layers present in the ANFIS model. In the first layer, input ranges are graded by applying the input membership functions to the input parameters. Then the weights are assigned to the ranged input variables in the second layer. In third step, the weighted input data are standardized. This operation is done while providing training to the ANFIS system itself. In the next layer the weighted input data from the third layer are shifted to pass through the defined rules. After this, the outputs obtained from the rules are noted and combinations of all such results are converted to the desired output value by using the systems' defuzzification method. In this study, the combination of least squares and backpropagation error method, that is, a hybrid algorithm, was used to train the ANFIS-based model. The results were analyzed using different types of membership functions, and the best ANFIS model topology having appreciable prediction capability of entry capacity of roundabout was established by calibrating different types of input membership function (Gaussian, constant, pi, trapezium, sigmoid, triangular and generalized bell shape), constant and linear type of output membership functions, defuzzification methods (weighted sum and weighted average), training cycles, and different number of input membership functions, shown in Table 3. Among the various ANFIS models, the best ANFIS model is based on highest coefficient of determination (R^2).

ANFIS Parameters used in the model development
No. of nodes = 4426
No. of linear parameters = 2187
No. of non-linear parameters = 63
Total no. of parameters = 2250
No. of training data pairs = 77
No. of testing data pairs = 33
No. of fuzzy rules = 2187
Optimization method: Hybrid1
Error tolerance = 0.01
Epochs = 20

Based on the above parameters for the development of the ANFIS model, the best ANFIS model was selected with the combination of Gaussian Bell input membership function, linear type output membership function, and weighted sum defuzzification method, having an R^2 value of 0.983. The limitation of the ANFIS model is that the model is capable of predicting values, but the equation-based model cannot be built up being a "black box" system (*14*).

Figure 2 gives the scatter plot between measured entry capacity and predicted entry capacity of MNLR and ANFIS models. It is found that these two models are statistically fit at the 95% confidence level under heterogeneous traffic conditions. The coefficient of correlation (R^2) is found to be 0.92, 0.93 and 0.98, 0.96 in both training and testing stage of MNLR and ANFIS model, respectively. This indicates that there is strong co-relation existing in the data sets ($R^2 > 0.80$). As described above, sometimes R^2 gives a biased estimate. Therefore, to assess the significance of these models and select the best model among them, MRI was applied.

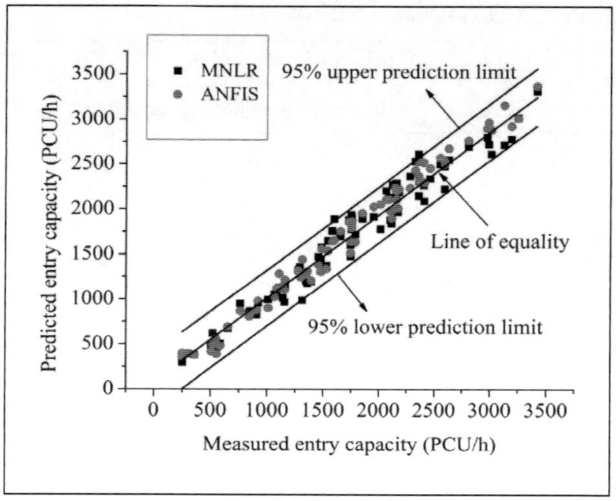

Figure 2. Scatter plot between measured and predicted entry capacity of MNLR and ANFIS model.

Significance Test of Developed Models. To assess the prediction performance of the two models MNLR and ANFIS, and to select the best model among them, MRI (*15*) was applied; details are given in Appendix 1. Subsequently, the models are ranked by using the following mathematical expressions given in Equation 2.

$$MRI = R_1 + R_2 + R_3 + R_4 + R_5 \quad (2)$$

R_1 = Ranking of a model based on best-fit calculations such as: coefficient of determination (R^2) and Nash–Sutcliffe model efficiency coefficient 'E'.

R_2 = Ranking of a model based on error measuring parameters such as average absolute error (AAE), maximum absolute error (MAE) and root mean square error (RMSE).

R_3 = Ranking of a model based on arithmetic calculations (mean 'μ' and standard deviation 'σ') of ratio of predicted and observed entry capacities.

R_4 = Ranking of a model based on 50% and 90% cumulative probability (P_{50} and P_{90}) values obtained from cumulative probability plot of C_P/C_O (ratio of predicted and observed entry capacity) values, and

R_5 = Ranking of a model based on prediction of entry capacity within ±20% accuracy level, calculated using the histogram and lognormal probability distribution of C_P/C_O (ratio of predicted and observed entry capacity) values.

Table 4. Sensitivity Analysis of MNLR Model

Sl. No.	Variables	Sensitivity (%)	Rank
1.	Weaving length (W_l)	17.45	3
2.	Critical gap (T_c)	18.43	1
3.	Diameter of central island (D)	6.00	7
4.	Weaving width (W_w)	18.34	2
5.	Entry width (E_w)	13.12	5
6.	Circulating flow (q_c)	16.00	4
7.	Follow-up gap (T_f)	10.62	6

From the literature, it is ascertained that when the predicted entry capacity is exactly equal to the field-observed entry capacity, the mean 'μ' and standard deviation 'σ' of C_P/C_O become 1 and 0, respectively. Therefore, these values closer to 1 and 0, respectively, indicate better performance of a predictive model. P_{50} value below and above 1 implies under and over prediction, respectively. The P_{90} parameter reflects the variation in C_P/C_O values for total observations. Thus, values of both P_{50} and P_{90} parameters closer to 1 indicate better performance of a predictive model. The MRI value as obtained for each entry capacity model developed in this study is shown in Appendix 1. It is observed that the MRI values are 5 and 13 in ANFIS (1st rank) and MNLR (2nd rank) models, respectively. Therefore it was found that the ANFIS model predicts the best performance in this study.

Sensitivity Analysis. In order to determine the percentage (%) contribution of each of the input variables in the model development, a sensitivity analysis was carried out. The sensitivity analysis of a given input is determined by varying the said input while other inputs are kept constant equal to their average value (*16*). From this analysis, the ANFIS model was considered to be the best model under heterogeneous traffic conditions. Owing to non-equation-based modeling (only forecast values), the practical applicability of the ANFIS model seems to be poor. The MNLR-based model has practical application because of its simplicity and manageability. The key point in the sensitivity analysis is that if the model has no expression for equation, percentage (%) sharing of each input variable cannot be determined. Therefore, the ranking of each input variable is determined in terms of the percentage contribution in the MNLR model development; details are given in Table 4.

It can be seen from Table 4 that critical gap, weaving width, and weaving length variables played major role in the model development. The percentage (%) contribution of these three variables is 18.43%, 18.34%, and 17.45% respectively. The above variables altogether contribute 54.22% in the model development. The critical gap variable was found to be the most prominent variable. This implies that the variable actually reflects driver behavior under local conditions.

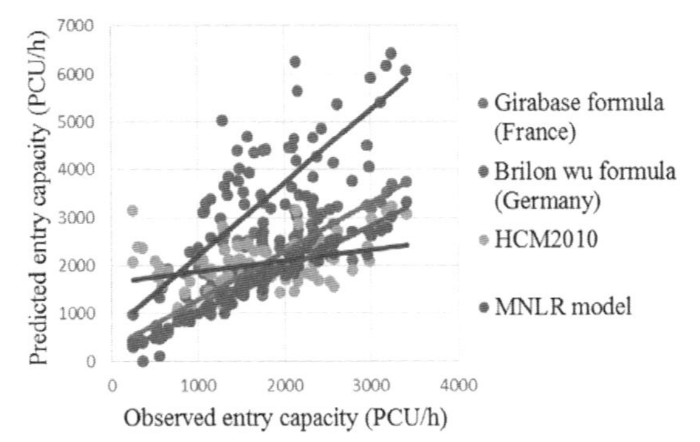

Figure 3. Comparison of roundabout entry capacity models.

Comparison with Other Roundabout Capacity Models. The proposed MNLR model is compared with existing capacity models such as the Girabase formula (France), Brilon wu formula (Germany), and HCM 2010 to assess the prediction ability under heterogeneous traffic conditions. The comparison is based on field-observed entry capacity with regard to predicted entry capacity by application of the above models; details are given in Figure 3.

It can be seen from the above analysis that Brilon wu formula (Germany) predicted higher capacity values in comparison with field-observed capacity values, with measured R^2 0.63. The prediction ranges are found to be between 1000 and 6500 PCU/h with corresponding field-observed capacity values of 250–3500 PCU/h. It was also found that the HCM 2010 model is poorly fitted under heterogeneous traffic conditions. The ranges are between 1800 and 2500 PCU/h with measured R^2 0.13. It is also noted that the model predicts lower capacity values in comparison with proposed the MNLR model.

The Girabase formula (France) shows similar capacity values, such as ranges between 400 and 3800 PCU/h with measured R^2 of 0.80. It is found that the proposed MNLR model predicts the capacity values within a range between 300 and 3300 PCU/h. The capacity values found from the proposed MNLR model are quite close to the field-observed capacity values with measured R^2 of 0.94, which is highest among all the existing capacity models. This indicates there is good correlation between data sets. The proposed model neither underestimates nor overestimates under heterogeneous traffic conditions.

Practical Applications

The results from statistical analysis of the MNLR model are useful for practical purposes. The model can be widely acceptable for its simplicity and manageability. The model seems to be workable on low to medium traffic volumes, and also it is useful for one to four-lane traffic volumes at un-signalized roundabouts. Sensitivity analysis is useful for planning and design because major contributing variables can be easily determined in the model development. Figure 3 gives an indication of the MNLR model in which the model may be applicable under heterogeneous traffic conditions in other developing countries.

Conclusions and Recommendations

- It is observed that the composition of traffic is widely heterogeneous in nature and varies from 0.26–61%, 10–79.5%, 14.55–79%, 0.05–33.1%, and 0.1–7% for HV, LMV, MC/S, BC, and HDV at selected roundabouts, respectively.
- To reflect the heterogeneous traffic, data are converted to static PCU as given on IRC: 65-1976 and gap-acceptance variables (critical gap and follow-up time) are determined by employing an INAGA method in this study. Inaccurate estimation of the critical gap gives a biased capacity value. The critical gap values were found to vary from 0.54 to 2.87 s, which is nearly half the value (4–4.6 s) of developed nations such as USA and European countries. These differences in critical gap values are attributable to the large number of two-wheelers in the present traffic condition. In fact, two-wheelers require marginal small gap as compared with other vehicles to merge into the major stream of traffic flow.
- From the two developed models of MNLR and ANFIS, the MRI reports that the ANFIS model is suitable for heterogeneous traffic conditions. It is ascertained that, from a statistical point of view, the ANFIS model holds good. But from practical point of view, the applicability of the model seems to be poor because the expression for equation cannot be built up. As compared with the ANFIS model, the MNLR model should be widely acceptable for its simplicity and manageability. Therefore the MNLR model is recommended for determining roundabout entry capacity under heterogeneous traffic conditions.
- The sensitivity analysis found that the critical gap variable is the most prominent variable for determining roundabout entry capacity. For accurate estimation of roundabout entry capacity, critical gap values should be carefully determined because this reflects local driver behavior. Comparison of models found that the HCM 2010 model is poorly

Appendix 1. Significance Test and Ranking of Developed Models

		Best-fit calculations			Error measuring parameters				Arithmetic calculations of P/O			Cumulative probability of P/O			± 20% accuracy (%)			Overall rank	
												Ratio* at							
Model	Data	R^2	E	R_1	AAE	MAE	RMSE	R_2	μ	σ	R_3	P_{50}	P_{90}	R_4	Log-normal	Histogram	R_5	MRI	Final rank
MNLR model	Training	0.92	0.91	2	165.0	663.49	216.48	2	0.99	0.13	2	0.97	1.19	2	87.47%	93.5%	2	8	2
	Testing	0.93	0.87		206.7	502.76	249.11		0.93	0.11		0.93	1.09		85%	88%			
ANFIS model	Training	0.98	0.98	1	84.21	265.11	103.45	1	1.00	0.11	1	0.99	1.07	1	91.48%	94.81%	1	5	1
	Testing	0.96	0.97		98.85	322.03	127.16		0.99	0.06		0.99	1.07		94.04%	100%			

Note: R_1, R_2, R_3, R_4, R_5 = ranking of variables in a model; *E* = Nash–Sutcliffe model efficiency coefficient; AAE = average absolute error; MAE = maximum absolute error; RMSE = root mean square error; C_P = predicted capacity; C_O = observed capacity; MRI = modified rank index.

fitted under heterogeneous traffic conditions. Indian traffic largely comprises two-wheelers rather than four-wheelers and large-sized vehicles. Two-wheelers forcibly create a gap to move into the circulating stream of traffic. Therefore, the critical gap values are randomly distributed, which leads to a violation of the principle of gap acceptance. But, the vehicles move in streamlined manner and the gaps are maintained consistently under homogeneous traffic flow conditions. Thus it is known that models developed under homogeneous traffic flow conditions are not fitted accurately under heterogeneous traffic conditions.

The proposed MNLR model provides ample information for practitioners, traffic engineers, and researchers as an effective tool for estimation of roundabout entry capacity with similar traffic conditions in other developing countries. Further research is recommended in two areas: (a) impact of cross-traffic for determining roundabout entry capacity, and (b) determination of dynamic PCU values to observe the effect of heterogeneity on roundabout entry capacity.

Author Contributions

The authors confirm contribution to the paper as follows: study conception and design: Ashish Kumar Patnaik, Ankit Raj Ranjan, Prasanta Kumar Bhuyan; data collection: Ashish Kumar Patnaik; analysis and interpretation of results: Ashish Kumar Patnaik, Ankit Raj Ranjan, Prasanta Kumar Bhuyan; draft manuscript preparation: Ashish Kumar Patnaik, Ankit Raj Ranjan, Prasanta Kumar Bhuyan. All authors reviewed the results and approved the final version of the manuscript.

References

1. Al-Masaeid, H., and M. Faddah. Capacity of Roundabouts in Jordan. *Transportation Research Record: Journal of the Transportation Research Board*, 1997. 1572: 76–85.
2. Patnaik, A. K., S. Rao, Y. Krishna, and P. K. Bhuyan. Empirical Capacity Model for Roundabouts Under Heterogeneous Traffic Flow Conditions. *Transportation Letters*, Vol. 9, No. 3, 2016, pp. 152-165. http://dx.doi.org/10.1080/19427867.2016.1203583
3. *Highway Capacity Manual*. TRB, National Research Council, Washington, D.C., 2000.
4. *Highway Capacity Manual 2010*. Transportation Research Board of the National Academies, Washington, D.C., 2010.
5. Rodegerdts, L., M. Blogg, E. Wemple, E. Myers, M. Kyte, M. Dixon, G. List, A. Flannery, R. Troutbeck, W. Brilon, N. Wu, B. Persaud, C. Lyon, D. Harkey, and D. Carter. *NCHRP Report 572:* Roundabouts in the United States. Transportation Research Board of the National Academies, Washington, D.C., 2007.
6. Gupta, A. K., P. Kumar, R. K. Sahoo, A. K. Sahu, and S. K. Sarangi. Performance Measurement of Plate Fin Heat Exchanger by Exploration: ANN, ANFIS, GA, and SA. *Journal of Computational Design and Engineering*, Vol. 4, No. 1, 2017, pp. 60–68. https://doi.org/10.1016/j.jcde.2016.07.002
7. Raff, M. S., and J. W. Hart. *A Volume Warrant for Urban Stop Signs*. Eno Foundation for Highway Traffic Control, Saugatuck, CT, 1950.
8. Wu, N. Equilibrium of Probabilities for Estimating Distribution Function of Critical Gaps at Unsignalized Intersections. *Transportation Research Record: Journal of the Transportation Research Board*, 2012. 2286: 49–55.
9. Patnaik, A. K, Y. Krishna, S. Rao, and P. K. Bhuyan. Development of Roundabout Entry Capacity Model Using INAGA Method for Heterogeneous Traffic Flow Conditions. *Arabian Journal for Science and Engineering*, Vol. 42, No. 9, 2017, pp. 4181–4199. http://dx.doi.org/10.1007/s13369-017-2677-x
10. Shafaei, R., M. Rabiee, and M. Mirzaeyan. An Adaptive Neuro Fuzzy Inference System for Makespan Estimation in Multiprocessor No-Wait Two Stage Flow Shop. *International Journal of Computer Integrated Manufacturing*, Vol. 24, No. 10, 2011, pp. 888–899. http://dx.doi.org/10.1080/0951192X.2011.597430
11. IRC 65. *Recommended Practice for Traffic Rotaries*. Indian Road Congress, 1976.
12. Arasan, V. T., and K. Krishnamurthy. Effect of Traffic Volume on PCU of Vehicles Under Heterogeneous Traffic Conditions. *Road Transportation Research ARRB*, Vol. 17, No. 1, 2008, pp. 32–49.
13. Kutner, M. H., J. Neter, C. J. Nachtsheim, and W. Li. *Applied Linear Statistical Models*. McGraw-Hill, New York, 2005.
14. Giustolisi, O., A. Doglioni, D. A. Savic, and B. W. Webb. A Multimodel Approach to Analysis of Environmental Phenomena. *Environmental Modelling and Software*, Vol. 22, No. 5, 2007, pp. 674–682. https://doi.org/10.1016/j.envsoft.2005.12.026
15. Beura, S. K., and P. K. Bhuyan. Modelling Quality of Bicycle Accommodations on Urban Road Segments Using Functional Networks and Multivariate Adaptive Regression Splines Techniques. Presented at 96th Annual Meeting of the Transportation Research Board, Washington, D.C., 2017.
16. Gandomi, A. H., G. J. Yun, and A. H. Alavi. An Evolutionary Approach for Modeling of Shear Strength of RC Deep Beams. *Materials and Structures*, Vol. 46, No. 12, 2013, pp. 2109–2119. https://doi.org/10.1617/s11527-013-0039-z

The Standing Committee on Highway Capacity and Quality of Service (AHB40) peer-reviewed this paper (18-04016).

Article

TRR
JOURNAL OF THE TRANSPORTATION RESEARCH BOARD

Transportation Research Record
2018, Vol. 2672(15) 44–54

DOI: 10.1177/0361198118798237
journals.sagepub.com/home/trr
SAGE

Estimating Passenger Car Equivalents on Level Freeway Segments Experiencing High Truck Percentages and Differential Average Speeds

Jianan Zhou[1], Laurence Rilett[2], Elizabeth Jones[1], and Yifeng Chen[3]

Abstract

In the *Highway Capacity Manual* (HCM), the passenger car equivalent (PCE) of a truck is used to account for the impacts of trucks on traffic flow. The 2010 HCM PCE values were estimated by the equal-density method using a FRESIM simulation. It was determined that the truck PCE for level freeway segments was 1.5 for all conditions. In the 2016 HCM, the PCE values were based on VISSIM simulation output at 1 min intervals along a three-lane, 13 mile (8 mile level and 5 mile graded) section of a roadway. It was determined that the truck PCE for level freeway segments was 2.0. It is hypothesized in this paper that the HCM PCE values are not appropriate for the western United States, which consistently experiences truck percentages higher than 25%, the maximum truck percentage published in the HCM PCE table. The HCM PCE procedure assumes that truck and passenger cars travel at the same average free-flow speed on level terrain. However, many heavy trucks in the western United States have speed limiters to improve fuel economy, and therefore travel slower than the speed limit. The interaction of high truck percentages and large speed differences may result in moving bottlenecks when trucks pass other trucks at low speed differentials. This may lead to an increased delay for the following passenger car vehicles. The 2016 HCM PCEs are based on three-lane simulations where the PCE is calculated based on near-capacity flows. This approach might not be appropriate for western states where these conditions rarely exist. This paper examines these effects using data from I-80 in western Nebraska. The paper develops new PCE values based on the 2010 HCM equal-density approach using calibrated CORSIM and VISSIM simulation models. It was found that the PCE values in the HCM 2010 and HCM 2016 underestimate the effect of heavy trucks on level terrain freeways that experience high truck percentages, and where different vehicle types have large differences in average free-flow speeds.

Trucks affect the quality of traffic flow and road capacity because of their size and operational characteristics. In the *Highway Capacity Manual* (HCM), the passenger car equivalent (PCE) of a truck, which represents the number of passenger cars that would have an equivalent effect on the traffic flow, is used to account for the presence of trucks (*1*). The PCEs for trucks on level freeway segments in the 2010 HCM are determined based on the "equal-density" method (*2*, *3*). The input for the model estimation was provided from a FRESIM simulation. It was determined that the truck PCE for level freeway segments was 1.5 for all conditions (*1*).

In the 2016 HCM, the PCE values are based on VISSIM simulation output at 1 min intervals along a three-lane, 13 mile (8 mile level and 5 mile graded) section of a roadway. The methodology employed an equivalency method in which an adjustment factor was identified that would make the 95th percentile mixed flow rate equal to the all-automobile 95th percentile flow rate (*4*). In other words, the PCE estimation procedure is "designed to estimate PCEs only for capacity flow" (*4*). A regression model was developed that estimates the PCE for a given FHWA truck type in which the independent variables were FHWA truck type (input as integer), the weight to horsepower ratio, the truck percentage, and the grade. Note that the truck type variable is a reflection of truck length, and only two of the 13 FHWA

[1]Department of Civil Engineering, University of Nebraska-Lincoln, Lincoln, NE
[2]Nebraska Transportation Center, University of Nebraska-Lincoln, Lincoln, NE
[3]Architecture, Engineering, Consulting, Operations, and Maintenance (AECOM), Southfield, MI

Corresponding Author:
Address correspondence to Jianan Zhou: jianan.zhou@huskers.unl.edu

truck types were modeled in the simulation (*5*). The equation also has "a limitation of not being able to evaluate PCE based on length of grade" (*5*). For level terrain the recommended 2016 HCM PCEs are disaggregated based on the percentage of single-unit trucks (e.g., FHWA classes 4–7) and heavy vehicles (e.g., FHWA classes 8–13), as well as the total truck percentage. A value of 2.0 was recommended for level terrain. When the traffic stream was 25% or more trucks on a 0% grade, the recommended PCE values ranged from 1.83 to 1.97. It is important to note, however, that the highest truck percentage value published in both the 2010 and 2016 HCM is 25%.

While the HCM PCE values may be appropriate for urban areas and large parts of the eastern United States, they may not be appropriate for the western rural United States, which consistently experiences high truck percentages. For example, data from the Nebraska Department of Transportation show that I-80 in western Nebraska consistently experiences truck percentages in the range of 25–60% (*6*). Given that it is estimated that truck freight tonnage will increase by 42% by 2040, this problem will continue into the future (*7*).

Another important assumption to note is that the HCM PCE procedure assumes that trucks and passenger cars travel at the same average free-flow speed on level terrain (*1*). However, many heavy trucks in the United States are governed through the use of speed limiters so that their maximum speeds are considerably less than the speed limit (*8*). This is done to optimize transportation costs by limiting fuel usage. Recently, the US government has recommended that speed limiters be applied to all heavy trucks to increase safety and reduce fuel usage (*9*).

Compounding the above issues is that the interaction of the high truck percentages and large speed differences may result in formation of moving bottlenecks when trucks pass other trucks at low speed differentials (e.g., a 67 mph truck passing a 66 mph truck). This may lead to increased delay for the following passenger car vehicles who, all else being equal, wish to travel faster. However, the impacts of the increased number and moving bottlenecks size experienced on level freeways are considered in neither the 2010 nor 2016 HCM PCE methodologies.

In addition, the recommended PCE values in the 2016 HCM are based on freeways with three lanes, where the PCE is calculated based on near-capacity flows. It is hypothesized that using a PCE estimation method that is based on conditions that rarely exist in the rural sections of western states might not be appropriate.

The objective of this study is to estimate and recommend PCEs for basic level freeway segments that experience high truck percentages and where heavy trucks and passenger cars have considerable differences in speed distribution. Empirical data from I-80 in western Nebraska were used to calibrate and validate the simulation models. The data from these simulation models were used to estimate PCEs following the PCE estimation methodology adopted in the 2010 HCM.

The paper is organized in the following manner. The literature on PCE estimation is first described, and then the PCE estimation methodology adopted in this paper is described in detail. The data collection, moving-bottlenecks/free vehicles identification, simulation model, and PCE determination are also presented. The section on moving-bottlenecks/free vehicles identification provides the definition and criteria for moving bottleneck identification on multilane freeways. The simulation model section provides details of the VISSIM and CORSIM models used in this research. The PCE estimation section describes the PCE estimation methodology and analyzes the PCE results. The PCEs are calculated under three conditions: single desired speed in CORSIM, single speed distribution in VISSIM, and empirical speed distributions in VISSIM. Lastly, an analysis of variance (ANOVA) analysis is conducted to identify the key factors (e.g., traffic volume and truck percentage) that influence the PCE values. The paper concludes with a discussion on future work.

Literature Review on PCE Estimation

The equal-impedance method defines the PCE as the number of passenger cars required to replace a truck in a mixed traffic stream such that the vehicles would experience the same impedance as a passenger-car-only traffic stream. The "impedance" metrics that have been used include the level of service (*10*), speed (*11*), density (*2*, *12*), and volume–capacity ratio (*13*). In the equal traffic volume or capacity method, the goal is to have the PCE-based passenger-car-only stream replicate the mixed traffic stream for specific traffic volume conditions (*14*). In some research the focus is on traffic volumes at capacity (e.g., LOS E) (*15*, *16*). It should be noted that this was the approach utilized in the 2016 HCM (*17*). Note that it is unclear to the authors how the 2016 HCM PCE values for different combinations of grade and grade lengths were obtained (*17*), given that the PCE regression equation developed as part of the underlying research does not have a variable related to grade length and so cannot estimate PCEs based on length of grade (*4*, *5*).

The headway-based PCE reflects spacing maintained by drivers in the proximity of trucks and cars, which attempts to measure the driver's freedom to maneuver (*18*). The headway-based PCE formulation is derived using Huber's equation and four headway values to estimate PCE values (*10*). The four headway values were car following car, car following truck, truck following car, and truck following truck headway. The formulation is sometimes simplified into the ratio of mean lagging headway for trucks to cars (*19*).

The queue–discharge–flow (QDF) method was used to estimate PCE values on a congested freeway (*20*). In this method, QDF capacity is considered the equivalent criterion because it governs the operation of a freeway after the onset of congestion (*21*).

The delay-based PCE is defined in such a way that if each truck in the traffic flow is replaced by a specific number of passenger cars, the replaced traffic flow will experience the same amount of delay (*22*). This method was initially developed for PCE at signalized intersections, and was later extended to determine PCE at work zones on a freeway (*23*).

The platoon-based method, initially proposed for a two-lane highway, is the only one that directly considers how moving bottlenecks impact PCE determination. In this method, a linear regression model is established between the number of followers and the volume of truck and passenger cars. The PCE is determined by the ratio of coefficients for trucks to passenger cars (*24*). However, the concept of a platoon has not been defined for four-lane freeways, and the potential correlation among independent variables may cause high standard errors for coefficients (*25*).

This paper estimates the truck PCEs using the equal-density method utilized in the 2010 HCM. The approach utilizes data from the CORSIM and VISSIM simulation models that have been calibrated using Nebraska empirical traffic flow data. The estimated PCE values are compared to both the 2010 and 2016 HCM values for level terrain and PCE values estimated in previous research.

2010 HCM Equal-Density Methodology for PCE Determination

The equal-density method is a specific type of equal-impedance method in which density is specified as the impedance (*2*, *8*). Figure 1 shows the process and key parameters. The goal is to identify the density for the subject traffic flow that provides the same density as the mixed and base flow situations.

It can be seen in Figure 1 that density–volume curves need to be developed for three different traffic flow types:

1. Base flow (passenger-car-only flow);
2. Mixed flow (traffic flow with passenger cars and a specific percentage of trucks); and
3. Subject flow (mixed traffic flow where a given number of passenger cars are replaced by an equal number of vehicles for which the PCE value is desired).

Based on previous research (*2*, *3*, *10*), the curves in Figure 1 are assumed to be quadratic in nature, as shown in Equation 1:

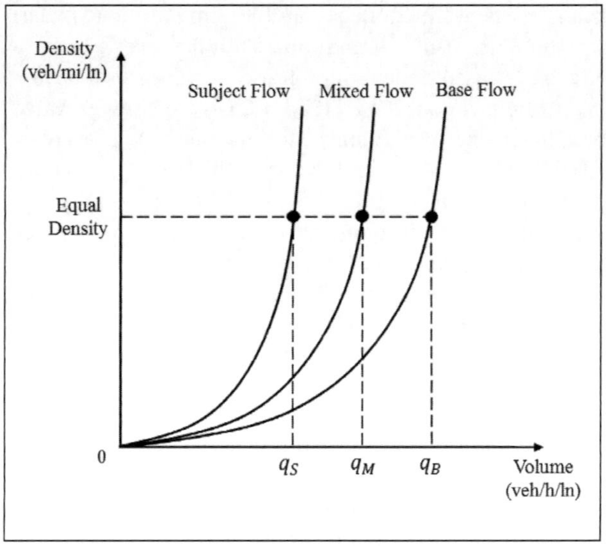

Figure 1. Required volume–density curves for estimating PCEs using the equal-density method.

$$D_i = a_{1i} \cdot V_i + a_{2i} \cdot V_i^2 \qquad (1)$$

where

i = indicator for traffic flow (i = 1, base flow; i = 2, mixed flow; i = 3, subject flow);

D_i = density estimated by 15 min based hourly volume and average speed for traffic flow i (veh/mi/ln);

V_i = hourly volume estimated by 15 min interval traffic volume for traffic flow i (veh/h/ln); and

a_{1i}, a_{2i} = estimated coefficients.

In this paper the subject vehicles were based on empirical data from Nebraska. A generic tractor–trailer and a generic single-axle unit truck were used to represent the Nebraska trucks. The average lengths and speed distributions of each truck type were based on empirical data. The percentage of passenger cars that are replaced is 5%, following 2010 HCM practice (*2*).

The need for large amounts of data to create the density–volume curves means the equal-density approach is typically based on simulation results. Each of the three traffic flows described above are simulated and the density values output. The coefficients of Equation 1 are then estimated using the density and volume data from the simulation data. Once the three density–volume curves are created, the truck equal-density PCE, denoted as ED-PCE in this paper, may be calculated at a given volume and truck percentage, as shown in Equation 2:

$$\mathrm{ED} - \mathrm{PCE}_{q_s, p_t} = \frac{1}{\Delta p}\left(\frac{q_B}{q_S} - \frac{q_B}{q_M}\right) + 1 \qquad (2)$$

where

$ED - PCE_{q_s,p_t}$ = equal-density-based PCE for trucks for given traffic flow volume q_S and truck percentage p_t;

Δp = percentage of subject vehicles (e.g., trucks) that replace an equivalent percentage of passenger cars in subject traffic flow (in this paper, this value is set to 5%);

q_B = base (e.g., passenger car only) flow volume that results in the same density as the given traffic flow (veh/h/ln);

q_M = mixed (e.g., p_t percent trucks and ($1 - p_t$) percent cars) flow volume that results in same density as the given traffic flow (veh/h/ln); and

q_S = subject (e.g., (p_t + 5) percent trucks and ($1 - p_t - 5$) percent cars) traffic flow volume (veh/h/ln).

Data Collection

Data were collected at 13 sites over a 222 mile segment of I-80 between mileposts 177 and 399 in Nebraska. The data collection occurred from June 1 through December 22, 2015. All data collection sites share the same geometric characteristics: level terrain (with grade 0–1%), two 12-foot lanes for each direction with a median, no horizontal curves, no vertical curves, and no work zones. All data were collected during daytime with clear weather and without strong winds. A total of 60 h of traffic flow data from all lanes in one direction for each site were collected using the Nebraska Transportation Center's (NTC) ITS van and video data collection equipment. The NTC ITS van is equipped with a video detection system and cameras mounted on a 42-foot telescope mast. At each location, the van was parked on the overpass above I-80 with cameras directed straight down on the lanes below.

Once the traffic stream video was obtained, it was further analyzed to obtain key traffic data. The Autoscope RackVision™ was used for the first step of data analysis (*26*). Virtual speed detectors were directly set on the video and traffic flow data were collected. The raw data included traffic volume, the time a given vehicle entered and exited the virtual detectors, vehicle speed, length, and type. Vehicles on the freeway were divided into five classes according to FHWA protocols (*27*): passenger car (including car with trailer, pick-up, and van), bus, single-unit truck, heavy truck, and recreational vehicles. Subsequently, the data were manually checked to ensure the vehicle types were correct. The Autoscope error rate on vehicle type identification was less than 5%. This is reduced to approximately 0% after the data clear step.

After the data reduction, a total of 48,903 valid vehicle records were available for analysis. The observed vehicles consisted mainly of passenger cars (70.2%) and trucks (29.1%). Very few buses and recreational vehicles were observed (only 0.7%). Of the trucks, 91% were classified as tractor–trailer/heavy trucks. The truck percentages were higher than 25% at 10 of 13 data collection sites. The empirical data also showed that the operating speeds of single-unit and heavy trucks are significantly lower than those of passenger cars. It should be noted that these sites would all be classified as operating under LOS A or B (*1*) using the 2010 HCM methodology, and LOS A, B, and C using the 2016 HCM methodology (*17*).

Free Vehicles Identification

In this research, free vehicles are defined as vehicles not in a moving bottleneck. A moving bottleneck is defined as a group of vehicles traveling on either the median or the shoulder lane in the same direction, and which influence the speed of one another. A vehicle is identified as being in a moving bottleneck based on (1) its leading and lagging headway, and (2) the critical headway (*28*, *29*). The detailed criterion for identifying the critical headway may be found elsewhere (*30*).

It was found that approximately 85% of vehicles may be classified as belonging to moving bottlenecks, and only 15% were identified as free-flow vehicles. The empirical speed distributions for free passenger cars, free single-unit, and free heavy trucks are shown in Figure 2. The figure shows the speeds of free passenger cars, single-unit, and heavy trucks are approximately normally distributed, and the average speeds of free single-unit and heavy trucks are, respectively, 7 and 10 mph slower than the average speeds of free passenger cars. These differences were found to be statistically significant using a standard *t*-test at the 95% confidence level (*31*).

Simulation Modeling Approach

In this paper, simulation models using both CORSIM 6.3 and VISSIM 9.0 were developed. To simulate the I-80 traffic flow conditions, the simulated network was designed as a four-link grid loop. Two of the links were 3.28 miles in length, and two of the links were 2.63 miles in length. Each link is one direction and has two 12 ft lanes and zero gradients. Each simulation run consists of the same three parts:

1. One-hour network loading so that the vehicles achieve a steady-state;
2. Two-hour steady-state with constant volume, where moving bottlenecks could be observed (these were the data used in the analysis); and
3. One-hour traffic unloading.

Four data collection points are chosen and set at equal distances on the network. At each data collection point, one detector is set for each lane.

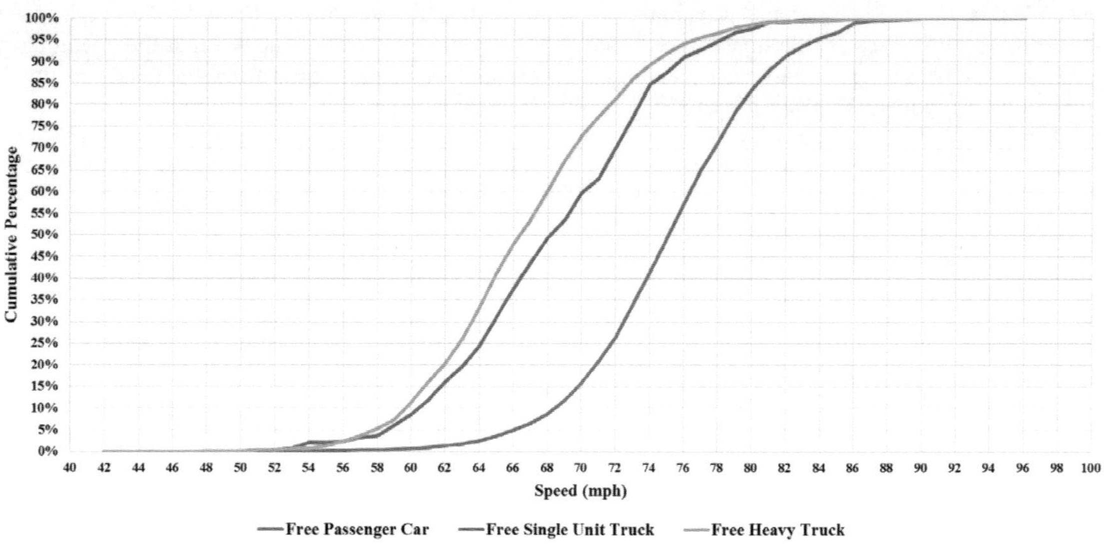

Figure 2. Empirical speed distribution for free vehicles.

The input vehicles are grouped into three classes: passenger cars, single-unit trucks, and heavy trucks. Here, the single-unit trucks represent trucks without trailers. These are modeled as 30–45 ft in length based on empirical observations. Single-unit trucks were set to 10% of the trucks based on the empirical analysis. Heavy trucks represent trucks with at least one trailer and three axles. They are modeled as being 65–100 feet in length based on empirical observations. They were modeled separately because, as shown in Figure 2, the mean speed of single-unit trucks was found to be different to those of passenger cars and heavy trucks, and this difference was found to be statistically significant using a standard *t*-test at the 95% confidence level (*31*).

To create the density–volume curves, several different combinations of volume and truck percentage were simulated. Six levels of traffic volume were simulated beginning at 500 veh/h/ln and increasing to 1,500 veh/h/ln in 200 veh/h/ln increments. Nine levels of truck percentage were simulated, beginning at 5%, and increasing to 85% at 10% increments. There were 20 runs for each combination of traffic volume and truck percentage, and each run had a different random number seed.

PCE Determination

There are five steps for calculating the ED-PCE values based on the 2010 HCM method and the output from the simulation model:

- Step 1: Develop the volume–density curve for the base (e.g., passenger car only) flow using simulation data with the six traffic volume levels and a fixed random seed. The hourly volume (veh/h/ln) is estimated based on the 15 min interval traffic volume, and the density (veh/min/ln) is estimated by dividing the 15 min based hourly volume by the average speed in the 15 min interval.
- Step 2: Develop the volume–density curve for the mixed (e.g., p_t percent trucks and $(1 - p_t)$ percent passenger cars) flow using simulation data with the same six traffic volume levels and the same random seed as in step 1. The hourly volume and density are estimated in the same manner as in step 1. Nine levels of truck percentage are simulated, as described above, and the mixed flow density–volume curves are estimated using Equation 1.
- Step 3: Develop the volume–density curve for the subject (e.g., $(p_t + 5)$ percent trucks and $(1 - p_t - 5)$ percent passenger cars) flow using data with the six traffic volume levels and the same random seed as in step 1. The hourly volume and density are estimated in the same manner as in step 1. Nine levels of truck percentage are simulated (e.g., 10–90% at 10% intervals). The subject flow density–volume curves are estimated using Equation 1.
- Step 4: The subject volume simulated ranges from 500 to 1,500 veh/h/ln at intervals of 200 veh/h/ln. For each subject flow volume (q_s), the corresponding equal-density value shown in Figure 1 is estimated from the volume–density curve for the subject flow. Then, the mixed volume (q_M) and

Table 1. Results for Average Equal-Density-Based PCEs

	Truck percentage								
Traffic volume (veh/h/ln)	5	15	25	35	45	55	65	75	85
Average equal-density-based PCEs with a single desired speed in CORSIM									
500	1.5	1.6	1.4	1.5	1.5	1.4	1.4	1.3	1.5
700	1.7	1.4	1.5	1.6	1.6	1.6	1.6	1.6	1.6
900	1.6	1.5	1.6	1.6	1.8	1.7	1.5	1.7	1.6
1,100	1.5	1.7	1.8	1.7	1.7	1.7	1.7	1.7	1.7
1,300	1.6	1.6	1.7	1.7	1.7	1.9	1.9	1.7	1.8
1,500	1.7	1.7	1.8	–	–	–	–	–	–
Average equal-density-based PCEs with a single speed distribution in VISSIM									
500	1.5	1.6	1.6	1.7	1.5	1.6	1.6	1.5	1.7
700	1.7	1.6	1.7	1.7	1.7	1.6	1.7	1.6	1.6
900	1.6	1.9	1.8	1.7	1.8	1.7	1.8	1.7	1.7
1,100	1.8	1.9	1.7	1.8	1.7	1.7	1.9	1.8	1.8
1,300	1.8	2.0	1.9	1.9	1.8	1.8	2.1	2.1	1.9
1,500	1.9	1.9	2.0	–	–	–	–	–	–
Average equal-density-based PCEs with empirical speed distributions in VISSIM									
500	2.2	2.3	2.5	2.4	2.5	2.5	2.4	2.3	2.5
700	2.4	2.3	2.4	2.6	2.4	2.3	2.5	2.5	2.5
900	2.5	2.5	2.7	2.5	2.8	2.5	2.6	2.7	2.6
1,100	2.7	2.5	2.8	2.8	2.7	2.7	2.7	2.8	2.5
1,300	2.8	2.8	2.9	2.7	2.8	2.8	2.8	2.7	2.7
1,500	2.8	2.7	2.7	–	–	–	–	–	–

Note: – = lacking adequate observations for this group.

base flow volume (q_B) that has the same equal density are estimated using the volume–density curves for the mixed and base flow, respectively.

- Step 5: The ED-PCE for each combination of volume (q_s) and truck percentage (p_t) for a specific random seed is estimated using Equation 2.
- Step 6: The above steps are repeated 19 times using different random number seeds following standard simulation protocols (*31*). A total of 20 ED-PCEs are obtained for each combination of q_s and p_t. These 20 PCE values are averaged to provide a final PCE estimate.

PCE Estimation Based on CORSIM

The reasons for using CORSIM are that: (1) it uses FRESIM as the microscopic simulation model representing traffic on highways and freeways; and (2) the 2010 HCM PCEs are calculated based on data simulated by FRESIM. The output data, including traffic volume and average speed, are aggregated by specific time intervals. One important factor to note is that CORSIM only allows vehicles to follow a single free-flow speed. This was not an issue for the 2010 HCM PCE estimation process, as it assumed that all vehicle types travel with the same average free-flow speed (*1*). As an aside, the 2016 HCM estimation process also made the same assumption for free-flow speed (*17*). As shown in Figure 2, this is clearly not true of I-80 in western Nebraska, and it is hypothesized that this is not true for much of the rural freeway sections in the western United States because of the widespread use of speed limiters (*9*).

The CORSIM ED-PCE mean values are shown in Table 1 as a function of traffic volume and truck percentage. It may be seen that the ED-PCE values range from 1.4 to 1.9, which are marginally higher than the 2010 HCM values. In general, as traffic volume increased, so did the ED-PCE values. There was no clear relationship observed between ED-PCE values and truck percentage.

Figure 3*a* shows the results graphically as a function of traffic volume and truck percentage. The larger rectangles represent the combined conditions while the smaller rectangles within it represent the values from one of the 20 repetitions. Not surprisingly, the results show that the PCE values obtained are all very similar to each other for a given run, and only marginally higher than the recommended 2010 HCM values.

As before, the results in Table 1 and Figure 3*a* are based on the assumption that all vehicle classes have the same single free-flow speed. It is impossible, using the current version of CORSIM, to model the effect of differential speed distributions that were observed in the field. Consequently, VISSIM was used to model this effect.

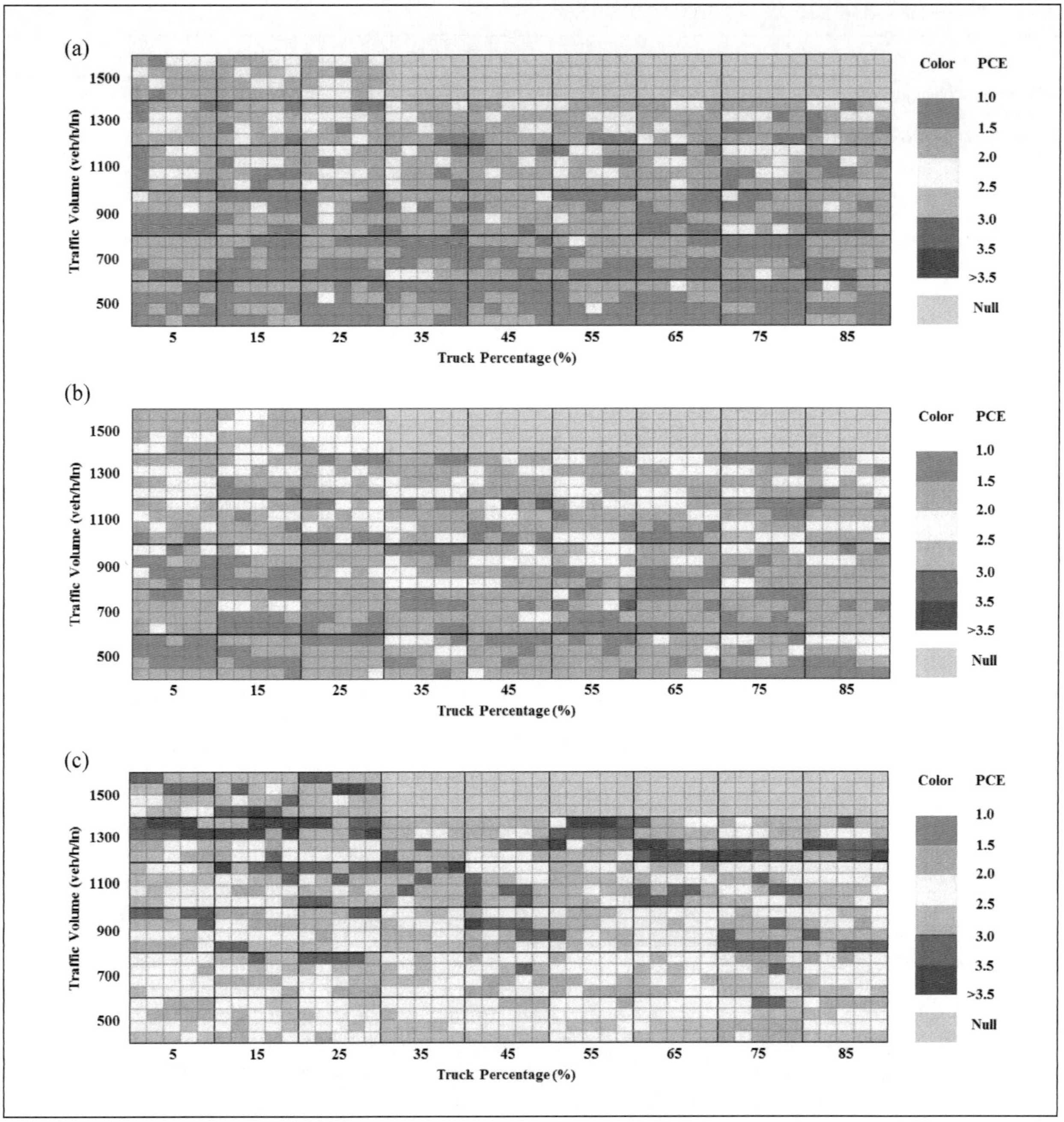

Figure 3. Equal-density-based PCEs with: (*a*) single desired speed in CORSIM; (*b*) single speed distribution in VISSIM; and (*c*) empirical speed distributions in VISSIM.

PCE Estimation Based on VISSIM

VISSIM allows different vehicle classes to follow different free-flow speed distributions and is therefore better suited to modeling operational conditions on western US roadways. The output data include detection of entering and exiting time, ordinal number, class, length, and speed for each vehicle. Note that two scenarios were examined. The first scenario used the same speed distribution for all vehicle types. This was done to see if VISSIM and CORSIM have the same assumption, whether similar PCE values could be obtained. The second scenario used the three empirical speed distributions shown in Figure 2. This was done to develop PCE values that are appropriate for the I-80 corridor in western Nebraska.

Table 2. Comparisons for Calculated ED-PCEs with HCM 2010 and HCM 2016

	Truck (%)								
	5	15	25	35	45	55	65	75	85
HCM 2010	1.5	1.5	1.5	1.5	1.5	1.5	1.5	1.5	1.5
HCM 2016 (30% SUT, 70% HT), 0% grade	2.30	2.04	1.97	1.97	1.97	1.97	1.97	1.97	1.97
HCM 2016 (50% SUT, 50% HT) 0% grade	2.31	2.02	1.93	1.93	1.93	1.93	1.93	1.93	1.93
HCM 2016 (70% SUT, 30% HT) 0% grade	2.12	1.89	1.83	1.83	1.83	1.83	1.83	1.83	1.83
Average ED-PCE (single free-flow speed in CORSIM)	1.6	1.6	1.6	1.6	1.7	1.7	1.6	1.6	1.6
Average ED-PCE (single speed distribution in VISSIM)	1.8	1.9	1.9	1.8	1.9	1.9	1.7	1.7	1.9
Average ED-PCE (empirical speed distributions in VISSIM)	2.6	2.5	2.7	2.6	2.6	2.6	2.6	2.6	2.6

Note: The estimated ED-PCEs in Table 1 were averaged so they could be compared with the values in HCM 2010 and 2016. The recommended PCE values on level terrain were 1.5 and 2.0 for the 2010 and 2016 HCM, respectively. SUT = single-unit truck; HT = heavy truck.

The VISSIM ED-PCE values, calculated using a single speed distribution, are shown in Table 1 and Figure 3*b* as a function of traffic volume and truck percentage. It may be seen that the ED-PCE values range from 1.5 to 2.1, which are marginally higher than the 2010 HCM values and similar to the values from the CORSIM analysis and the 2016 HCM value for level terrain. As before, as traffic volume increased, so did the ED-PCE values. There was no clear relationship observed between the ED-PCE values and truck percentage. These results indicate that the VISSIM and CORSIM simulation models give similar ED-PCE results, if a single free-flow speed is used in CORSIM and a single speed distribution is used in VISSIM.

The VISSIM ED-PCE values, calculated using the three empirical speed distributions are shown in Table 1 and Figure 3*c* as a function of traffic volume and truck percentage. It may be seen that the ED-PCE values range from 2.2 to 3.0, which are approximately 50–100% higher than the 2010 HCM values and approximately 15–50% higher than the HCM value for level terrain. As before, as traffic volume increased so did the ED-PCE values. It is hypothesized that the speed differential causes moving bottlenecks to form similar to what was observed in the field. This translated into higher density values for the traffic for a given truck volume and resulted in higher PCE values. There was no clear relationship observed between ED-PCE values and truck percentage. This result is similar to what was found in the 2010 HCM. In contrast, the PCEs in the 2016 HCM decreased with truck percentage. It is hypothesized that this occurred because the 2016 HCM PCE values were based on near-capacity conditions.

ANOVA Analysis

To explore the influence of factors, an ANOVA analysis was conducted. It was found that with respect to the ED-PCEs values:

1. The ED-PCEs increased with traffic volume for all scenarios;
2. The ED-PCEs did not vary appreciably with truck percentage for all scenarios;
3. There was no difference between ED-PCEs calculated by VISSIM data with a single free-flow speed distribution and by CORSIM data with single free-flow speed; and
4. The ED-PCEs calculated using VISSIM data with the empirical speed distributions were higher than that calculated using either VISSIM data or CORSIM data with a single free-flow speed distribution.

The above results were all statistically significant based on a *F*-test at the 95% level of confidence (*31*).

Discussion

A summary of the ED-PCEs estimated in this paper with the HCM 2010 PCE values is shown in Table 2. The table shows that the calculated values in this study, when it is assumed that trucks and passenger cars travel at the same speed, are only marginally higher than the 2010 HCM values. However, when the empirical speed distributions are used, the estimated ED-PCE values are considerably higher. Also shown in Table 2 are the HCM 2016 PCE values. It may be seen that for truck percentages higher than 25%, the ED-PCE values in this paper are 30% higher. In addition, the ED-PCE values are approximately 20–70% higher than those found in previous research for high truck percentages and low volumes (*2, 11–18*).

It is argued in this paper that the higher ED-PCE values better capture the relationship between heavy vehicles and passenger cars on this corridor. For each of the data collection sites it was found that the observed speeds were 0–20% lower than those which were to be expected from the HCM, if the speed–flow curve for free-flow speed =

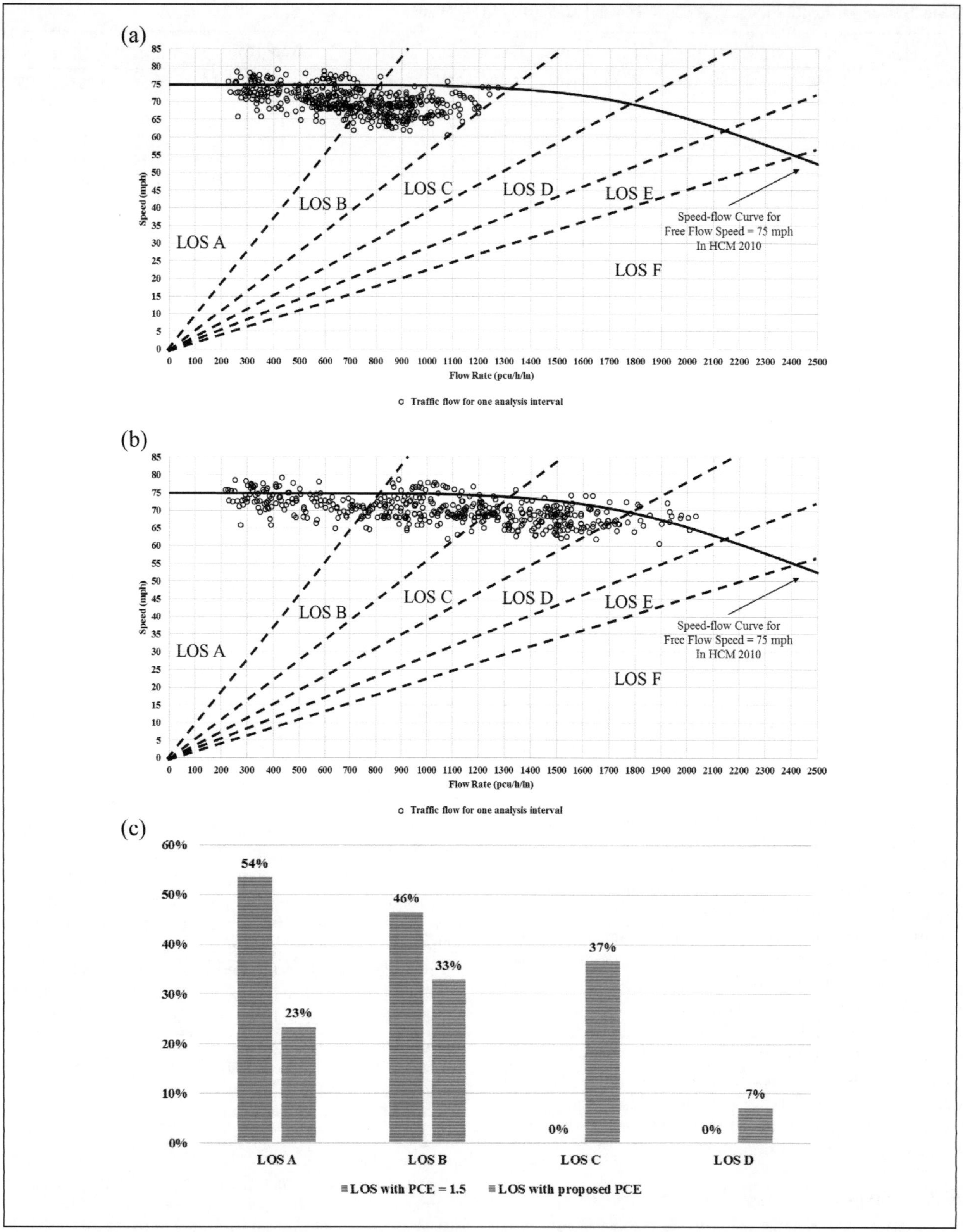

Figure 4. Relationship between observed traffic flow and free-flow speed = 75 mph speed–flow curve in HCM 2010 with (*a*) PCE = 1.5; (*b*) proposed PCEs; and (*c*) comparison of LOS results between using PCE = 1.5 and proposed PCEs.

75 mph (exhibit 11-2 in HCM 2010 [*2*]) was selected for analysis, as shown in Figure 4*a*. It may be seen in Figure 4*a* that the observed speeds correspond to the LOS C or D section on the curve.

If the larger PCEs proposed in this research are used, the estimated LOS will change, as shown in Figure 4*b*. As shown in Figure 4*c*, if the 2010 HCM PCE value is used, 54% of the sites would be classified as having a LOS A and 46% of the sites would be classified as having a LOS B. However, if the proposed PCEs are used, the results would be 23% LOS A, 33% LOS B, 37% LOS C, and 7% LOS D.

It is hypothesized that because of the presence of the large percentage of moving bottlenecks, the HCM analysis might not capture reality because it tends to assume that traffic is uniformly distributed at low volumes. For example, the PCEs for the 2016 HCM were based on simulations of three-lane roads where passing would be much easier at low volumes and moving bottleneck formation would be much lower.

Concluding Remarks

This research estimated truck PCEs for level freeways that experience high truck percentages and large speed differentials between heavy trucks and passenger cars. It was found that the ED-PCE values were considerably higher than those published in the 2010 HCM and the 2016 HCM for level terrain (*2*). It is hypothesized that the results found in this paper better capture the interaction between trucks and passenger cars that have differing free-flow speeds because it explicitly captures the moving bottleneck formation and the resulting delay and localized congestion.

It should be noted that the results are based on empirical data obtained from I-80 in western Nebraska. Obviously, the results may change if data from other western US locations are used to calibrate the model. However, it is hypothesized that the trend to larger numbers of heavy trucks on the US freeway system, coupled with the use of speed limiters, will only increase in the future. Consequently, new PCE values are required if transportation agencies are to plan and operate their transportation systems under this new paradigm. This research was conducted on freeways with two-lane segments. It would be advantageous to conduct research on freeways with more than two lanes in one direction, and in rolling and mountainous terrain where the effects may be even more pronounced. It is feasible to estimate PCEs for other roadway geometrics and terrain as they may be readily modeled in the microsimulation models. However, it is critical that these microsimulation models are calibrated for the conditions to be studied. It would be useful to compare the PCE estimation approach outline in this paper with the PCE estimation equal-capacity approach advocated in the 2016 HCM. As stated earlier, these results are particularly important for departments of transportation in the western United States. It is hypothesized that they could improve the capacity of their rural freeways by: (1) adding a new lane on certain sections; (2) restricting trucks to the shoulder lane; or (3) adjusting the speed limits for all vehicles such that the speed distributions are similar.

Acknowledgments

This paper is supported by the Nebraska Department of Transportation through the Research Project (NTRC No. TSPT1-02) *Assessing Passenger Car Equivalency Factors for High Truck Percentage*, and this support is gratefully acknowledged.

Author Contributions

The authors confirm contribution to the paper as follows: study conception and design: LR, EJ, JZ, YC; data collection: LR, EJ, JZ, YC; analysis and interpretation of results: LR, EJ, JZ, YC; draft manuscript preparation: LR, EJ, JZ, YC. All authors reviewed the results and approved the final version of the manuscript.

References

1. *Highway Capacity Manual 2010*. Transportation Research Board of the National Academies, Washington, D.C., 2010.
2. Webster, N., and L. Elefteriadou. A Simulation Study of Truck Passenger Car Equivalents (PCE) on Basic Freeway Sections. *Transportation Research Part B: Methodological*, Vol. 33, No. 5, 1999, pp. 323–336.
3. Sumner, R., D. Hill, and S. Shapiro. Segment Passenger Car Equivalent Values for Cost Allocation on Urban Arterial Roads. *Transportation Research Part A: General*, Vol. 18, No. 5, 1984, pp. 399–406.
4. Dowling, R., G. List, B. Yang, E. Witzke, and A. Flannery. *NCFRP Report 31: Incorporating Truck Analysis into the Highway Capacity Manual*. Transportation Research Board of the National Academies, Washington, D.C., 2014.
5. Yang, B. *On the HCM's Treatment of Trucks on Freeways*. Master's thesis. North Carolina State University, Raleigh, NC, 2013.
6. *Traffic Flow Map of the State Highways, State of Nebraska, 2015*. U.S. Department of Transportation, Washington, D.C., 2015.
7. Mendez, V. M., C. A. Monje, Jr, and V. White. *Beyond Traffic: Trends and Choices 2045—A National Dialogue About Future Transportation Opportunities and Challenges in Disrupting Mobility*. Springer International Publishing, New York City, NY, 2017.
8. Bishop, R. *CTBSSP Synthesis 16: Safety Impacts of Speed Limiter Device Installations on Commercial Trucks and Buses*. Transportation Research Board of the National Academies, Washington, D.C., 2008.

9. Rosekind, M. R., and T. S. Darling, III. *Notice of Proposed Rulemaking; Federal Motor Vehicle Safety Standards; Federal Motor Carrier Safety Regulations; Parts and Accessories Necessary for Safe Operation; Speed Limiting Devices*. Docket No. NHTSA-2016-0087, Docket No. FMCSA-2014-0083, 2016.
10. Huber, M. J. Estimation of Passenger-Car Equivalents of Trucks in Traffic Stream. *Transportation Research Record: Journal of the Transportation Research Board*, 1982. 869: 60–70.
11. Elefteriadou, L., D. Torbic, and N. Webster. Development of Passenger Car Equivalents for Freeways, Two-lane Highways, and Arterials. *Transportation Research Record: Journal of the Transportation Research Board*, 1997. 1572: 51–58.
12. Rakha, H., A. Ingle, K. Hancock, and A. Al-Kaisy. Estimating Truck Equivalencies for Freeway Sections. *Transportation Research Record: Journal of the Transportation Research Board*, 2007. 2027: 73–84.
13. Okura, I., and N. Sthapit. Passenger Car Equivalents of Heavy Vehicles for Uncongested Motorway Traffic from Macroscopic Approach. *Journal of Infrastructure Planning and Management*, Vol. 512(IV-27), 1995, pp. 73–82.
14. Alecsandru, C., S. Ishak, and Y. Qi. Passenger Car Equivalents of Trucks on Four-lane Rural Freeways under Lane Restriction and Different Traffic Conditions. *Canadian Journal of Civil Engineering*, Vol. 39, No. 10, 2012, pp. 1145–1155.
15. Fan, H. S. Passenger Car Equivalents for Vehicles on Singapore Expressways. *Transportation Research Part A: General*, Vol. 24, No. 5, 1990, pp. 391–396.
16. Yeung, J. S., Y. D. Wong, and J. R. Secadiningrat. Lane-Harmonised Passenger Car Equivalents for Heterogeneous Expressway Traffic. *Transportation Research Part A: Policy and Practice*, Vol. 78, 2015, pp. 361–370.
17. *Highway Capacity Manual: A Guide for Multimodal Mobility Analysis*, 6th ed. Transportation Research Board, Washington, D.C., 2016.
18. Krammes, R. A., and K. W. Crowley. Passenger Car Equivalents for Trucks on Level Freeway Segments. *Transportation Research Record: Journal of the Transportation Research Board*, 1986. 1091: 10–17.
19. Okura, I., and N. Sthapit. Microscopic Headway Method of Estimating Passenger Car Equivalents. *Proceedings of Infrastructure Planning*, Vol. 17, 1995.
20. Al-Kaisy, A., Y. Jung, and H. Rakha. Developing Passenger Car Equivalency Factors for Heavy Vehicles During Congestion. *Journal of Transportation Engineering*, Vol. 131, No. 7, 2005, pp. 514–523.
21. Al-Kaisy, A. F., F. L. Hall, and E. S. Reisman. Developing Passenger Car Equivalents for Heavy Vehicles on Freeways During Queue Discharge Flow. *Transportation Research Part A: Policy and Practice*, Vol. 36, No. 8, 2002, pp.725–742.
22. Benekohal, R. F., and W. Zhao. Delay-Based Passenger Car Equivalents for Trucks at Signalized Intersections. *Transportation Research Part A: Policy and Practice*, Vol. 34, No. 6, 2000, pp. 437–457.
23. Chitturi, M. V., and R. F. Benekohal. Effect of Work Zone Length and Speed Difference Between Vehicle Types on Delay-Based Passenger Car Equivalents in Work Zones. Presented at 87th Annual Meeting of the Transportation Research Board, Washington, D.C., 2008.
24. Van Aerde, M., and S. Yagar. Capacity, Speed and Platooning Vehicle Equivalents for Two-Lane Rural Highways. *Transportation Research Record: Journal of the Transportation Research Board*, 1984. 971: 58–67.
25. Gunst, R. F., and J. T. Webster. Regression Analysis and Problems of Multicollinearity. *Communications in Statistics: Theory and Methods*, Vol. 4, No. 3, 1975, pp. 277–292.
26. Yu, X., P. Prevedouros, and G. Sulijoadikusumo. Evaluation of Autoscope, SmartSensor HD, and Infra-Red Traffic Logger for Vehicle Classification. *Transportation Research Record: Journal of the Transportation Research Board*, 2010. 2160: 77–86.
27. James, L., and P. E. Randall. *Traffic Recorder Instruction Manual*. Texas Department of Transportation, Austin, TX, 2012.
28. Fitzpatrick, K., S. P. Miaou, M. Brewer, P. Carlson, and M. D. Wooldridge. Exploration of the Relationships Between Operating Speed and Roadway Features on Tangent Sections. *Journal of Transportation Engineering*, Vol. 131, No. 4, 2005, pp. 261–269.
29. Al-Kaisy, A., and S. Karjala. Car-Following Interaction and the Definition of Free-Moving Vehicles on Two-Lane Rural Highways. *Journal of Transportation Engineering*, Vol. 136, No. 10, 2010, pp. 925–931.
30. Zhou, J., L. R. Rilett, and E. Jones. *Assessing Passenger Car Equivalency Factors for High Truck Percentages*. Nebraska Department of Roads, Lincoln, NE, 2017.
31. Spiegelman, C., E. S. Park, and L. R. Rilett. *Transportation Statistics and Microsimulation*. CRC Press, Boca Raton, FL, 2010.

The Standing Committee on Highway Capacity and Quality of Service (AHB40) peer-reviewed this paper (18-05573).

The contents of this paper reflect the views of the authors, who are responsible for the facts and the accuracy of the information presented here, and are not necessarily representative of the Nebraska Department of Transportation.

Article

TRR
JOURNAL OF THE TRANSPORTATION RESEARCH BOARD

Transportation Research Record
2018, Vol. 2672(15) 55–65

DOI: 10.1177/0361198118777407
journals.sagepub.com/home/trr
SAGE

Saturation Flow Model for Signalized Intersection under Mixed Traffic Condition

Sabyasachi Biswas[1], Souvik Chakraborty[2], Indrajit Ghosh[3], and Satish Chandra[4]

Abstract

Saturation flow is one of the most important functional parameters at signalized intersections. It is to be noted that saturation flow is a functional measure of the intersection operation, which indicates the probable capacity if working in an ideal situation. However, determination of the saturation flow is a challenging task in developing countries like India where vehicles with diverse static and dynamic characteristics use the same carriageway. At the same time, it is influenced by several other factors. In this context, the present research is carried out to examine the effects of traffic composition, approach width and right-turning movements on saturation flow under heterogeneous traffic conditions. This paper proposes a model for computing saturation flow at the signalized intersection under mixed traffic condition based on Kriging approach. A detailed comparison of the mean saturation flow values obtained by the conventional method, regression method, and Kriging method has also been presented. Low mean absolute percentage error values (<5%) have been obtained for saturation flow by Kriging method with respect to the conventional method. Finally, the proposed models are used to evaluate the impact of right-turning vehicles on saturation flow under shared lane condition.

Signalized intersections are one of the major influencing elements of a road network for designing an efficient urban roadway. In this regard, saturation flow is a crucial factor in designing the signal cycle and the corresponding green phase at the intersections. Saturation flow is quantified in terms of vehicles per hour of green under homogeneous traffic condition. The traditional method for estimating saturation flow as described in the Highway Capacity Manual (HCM) (*1*), based on various multiplicative factors, is used to adjust the estimated saturation flow under prevailing circumstances. These adjustment factors are meant for various influencing parameters. Such a setup, although appropriate for homogeneous traffic condition, is not directly applicable for highly heterogeneous traffic condition.

Initial studies on estimation of saturation flow are based on the headway ratio method proposed by Greenshields et al. (*2*) in 1947. The primary idea of this method is to estimate the saturation flow based on saturation headway at the signalized intersections. The method assumes that the headway between the vehicles becomes constant after passing of the first four vehicles in the queue. This is done to avoid vehicles' inertia, at the beginning, after the onset of green phase (*3*). A detailed account of this method is also available in the special report published as part of the HCM by the National Research Council (*4*). There are many studies available in the literature on measurement of saturation flow based on headway method (*5–9*). Webster and Cobbe (*8*) established a saturation flow model based on road width. Similarly, IRC 93-1994 (*10*) estimates saturation flow based on approach road width, although, such an ideal scenario rarely exists in practice. Additionally, some adjustment factors corresponding to turning movements and proportion of heavy vehicles have also been proposed. However, this model does not consider the effect of various vehicle categories: for example, motorized two-wheeler has an impact on saturation flow due to its small size and better maneuverability. At the same time, the saturation flow of the intersection varies among

[1]Amity University Noida, Noida, India
[2]University of Notre Dame, Notre Dame, IN
[3]Indian Institute of Technology (IIT) Roorkee, Uttarakhand, India
[4]CSIR-Central Road Research Institute (CRRI), New Delhi, India

Corresponding Author:
Address correspondence to Sabyasachi Biswas:
sabyasachibiswas01@gmail.com

different cycles under this non-lane discipline traffic condition. Hence existing methods may not be appropriate for a highly heterogeneous traffic situation.

A number of researchers have developed saturation flow models based on prevailing traffic condition for homogeneous traffic situations. However, a limited number of studies (*11–13*) has been carried out on saturation flow under heterogeneous condition. A few studies considered that the drivers follow a low level of lane discipline, though this is not reflected in reality. In TRL method (*14*), saturation flow is computed as the number of vehicles in the saturated green time. Turner and Harahap (*15*) contributed significantly toward the advancement of TRL method. The regression technique (*16*, *17*) involves developing models relating various influential parameters to the saturation flow. To this end, it is essential to identify the significant parameters influencing the saturation flow. Most of the studies carried out to date are based on traffic proportion, and road width of the intersection approaches. Branston et al. (*17*) presented two multiple linear regressions methods known as 'asynchronous' and 'synchronous.' Saturated green time, the vehicle numbers in various type, and lost time were considered as independent factors to develop the saturation flow model. The study further explores the temporal impact on saturation flow. The study observed higher saturation flow during peak hours. The saturation flow at daytime peak hour was found to be more compared with the evening peak time. However, one major drawback of the regression method is its accuracy. Regression-based saturation flow models are mostly linear with very few parameters. Such models rarely yield accurate results for highly heterogeneous conditions. Hence, there is an urgent need to develop saturation flow model for highly heterogeneous traffic conditions at signalized intersections. On the other hand, the presence of right-turning vehicles (for left-hand drive) at signalized intersections tends to cause delay and chances of an accident. It reduces the intersection's capacity. Exclusive right-turn lanes and different phasing strategies (e.g., permitted, protected, and protected-permitted) are employed in many countries to manage right-turning traffic. In the literature, it was found that various methods were established in different countries (*1*, *4*, *18–21*) to obtain the adjustment of saturation flow for right-turning movement. U.S. HCM (*1*), the Swedish approach (*19*), and Australian Road Research Board proposed methods for adjustment factor for turning movements which are based on turning vehicle proportion only. The traffic composition is not considered for these adjustment factors. However, the Canadian Capacity Guide (*18*) established saturation flow adjustment factors based on vehicle composition as well as turning vehicle proportion at a shared lane. Under such circumstances, for practical implementation of saturation flow model, it is necessary to determine a suitable adjustment factor for (*22*) shared right-turning and through vehicles (F_{rt}) under highly heterogeneous traffic condition.

In an attempt to overcome the above-mentioned issues, Kriging-based saturation flow model has been proposed in this study. Unlike previous studies, both the road width and classified vehicle counts have been considered as influential parameters. In addition, this paper presents a novel approach for estimation of saturation flow adjustment factor due to the right-turning movements at shared lane condition. Compared to conventional regression techniques, Kriging has the following advantages:

- The functional form of Kriging consists of two portions, one is the polynomial trend and the other is Gaussian process. The Gaussian process portion minimizes the local variance based on an autocorrelation function. This results in an improved prediction accuracy as compared with traditional regression approaches, such as least squares.
- Kriging presents a bi-level approximation methodology, initiated at a global level by utilizing the trend function (polynomial part) and secondly at a local level based on appropriate covariance function. Therefore, predicted results obtained by Kriging approach are more accurate compared with regression-based methods.
- Kriging techniques provide the best linear unbiased method (*23*).
- Significantly small data sets are required to develop a Kriging model compared with other models (*23*). Therefore, Kriging technique is generally used for statistical modeling in the field of reliability (*23*, *24*), uncertainty quantification (*23*), and optimization (*24*).

However, use of Kriging in traffic engineering is quite rare. To the best of our knowledge, this is the first attempt where Kriging has been utilized to model saturation flow. Moreover, unlike previous studies, both vehicle composition and road width have been incorporated into the developed model. Subsequent sections describe the proposed approach.

Background Study

Estimation of Passenger Car Unit (PCU)

The PCU values are estimated by various methods, mostly for heavy vehicles or buses (*25*, *26*) which are the other vehicle types excluding passenger cars. Consequently, the concept of PCU is incorporated into the saturation flow model for heterogeneous traffic condition. PCU values allow different vehicle types within the traffic flow group to be assessed in a consistent manner. In the present study, PCU is computed as the headway ratio of the two vehicles as

$$(\text{PCU})_i = \frac{\bar{h}_{i-i}}{\bar{h}_{c-c}} \tag{1}$$

where
$(\text{PCU})_i$ = PCU of the *i*th vehicle category;
$\bar{h}_{i-i}$ = mean headway of a type *i* vehicle followed by a type *i* vehicle; and
$\bar{h}_{c-c}$ = mean headway of a car followed by a car.

It requires observation of headway combination while both leading and the following vehicles are of the same category. Though, it can be seen that it is quite difficult to estimate the PCU for each vehicle category due to the lack of headway combinations. Therefore, several hours of data have been extracted from video to obtain more headway combinations (*27*).

Estimation of Saturation Flow

Saturation flow rate of a signalized intersection is defined as the number of vehicles that would cross the stop line, given the signal is green. It is evident from the definition that the saturation flow is expressed as vehicles per hour of green (*1*). In conventional method, incorporating PCU, the saturation flow of an intersection under highly heterogeneous traffic condition is computed as (*13*, *18*, *28*)

$$s_b = \left[\sum_i n_{ni}\text{PCU}_i\right]\frac{3600}{g_s} \tag{2}$$

where
s_b = saturation flow per standard approach width of single lane (w_b) in terms of PCU/h/lane;
PCU_i = PCU for *i*-vehicle category; and
g_s represents saturated green period.
Saturation flow for entire approach width (s) in terms of PCU/h is evaluated by the conventional method based on PCU values.

However, such a representation ignores the effect of width. Ideally, saturation flow at an intersection should be a linear function of width and represented as

$$s = s_b \frac{w}{w_b} \tag{3}$$

where, s is the saturation flow at the intersection in terms of PCU/h. Though, such an ideal scenario rarely exists in practice. Additionally, the common limitation of Indian Standard (*10*) inhabits in the hypothesis that saturation flow has a linear relationship with approach width.

Methodology

Development of a saturation flow model is a three-step procedure. The first step involves selection of appropriate site and collection of data. The second step involves the development of a model by utilizing the collected field data. The final step involves the adjustment of the saturation flow for road width and turning vehicles. In this section, the three steps involved are discussed. First, details regarding data collection have been provided. Then, formulations of the proposed models have been discussed.

Data Collection

Traffic data were collected from seventeen approaches of different intersections located in New Delhi and Noida, India. All the intersection selections were based on the following criteria:

- Approximately level gradient approach road
- Different approach speed limits
- Variation of traffic volumes and approach width
- Fixed time signal
- No bus stop in exit point of the approach roads
- No parking in the approach, upstream or downstream

Among various factors that could affect the saturation flow under mixed traffic condition, the present study focuses on the influence of traffic composition, approach road width and right-turning movements in the traffic stream. From the prior studies, it has been observed that these three factors are the factors which most influence the saturation flow at signalized intersections. In the absence of any thorough study for finding the effect of these individual influencing factors on saturation flow, Indian standards still follow the U.S. HCM or UK guidelines. This study is the first step in this direction to come up with the adjustment factors which are applicable for mixed traffic condition.

Video recording was the primary technique to capture traffic flow followed by data extraction. Data at only one approach, for each of the selected intersections, were collected. Two cameras were utilized for this purpose. The first camera was mounted on a vantage point adjacent to the intersection and traffic approaching the intersection along the selected approach was recorded. It was ensured that this camera should record traffic data at least 100 m upstream from the intersection stop line. The second camera recorded the signal indications displayed to the traffic for the same approach. Since the focus of

Table 1. Key Characteristics of the Intersections

Approach ID	Name of the intersection	Target movements	Approach**	Approach width (m)	No. of cycles observed	Cycle length/ green time (s)	Composition (%) (C: BC: 3W: 2W:HV)
A1	Noida sector 5–6 intersection	Through and right-turning	EB	8	24	170/25	32:8:4:54:1
A2	Noida sector 5–6 intersection		WB	9.2	22	170/25	27:7:3:59:3
A3	Noida sector 5–6 intersection		NB	10.2	24	170/50	47:11:5:29:7
A4	Noida sector 5–6 intersection		SB	9.7	23	170/50	49:8:5:28:11
A5	Noida Sector 6–7 intersection		NB	10.7	23	200/60	41:8:6:33:10
A6	Noida Sector 6–7 intersection		SB	10.9	23	200/60	44:7:5:35:9
A7	Prithviraj Road and Rajesh Pilot Marg intersection		EB	9	25	160/30	37:7:15:36:6
A8	Prithviraj Road and Rajesh Pilot Marg intersection		WB	9	25	160/30	43:9:18:28:3
A9	Prithviraj Road and Rajesh Pilot Marg intersection		SB	6	25	160/30	46:13:11:27:3
A10	Dr. Zakir Hussain Marg and Subramaniam Bharti intersection		EB	9	29	200/45	41:11:23:24:1
A11	Dr. Zakir Hussain Marg and Subramaniam Bharti intersection		WB	9	28	200/45	43:10:19:27:1
A12	Dr. Zakir Hussain Marg and Subramaniam Bharti intersection		NB	9	29	200/45	52:10:17:19:2
A13	Dr. Zakir Hussain Marg and Subramaniam Bharti intersection		SB	9	29	200/45	54:12:13:19:1
A14	Pruthvi Raj Chowk 2nd intersection		EB	10.2	24	270/80	42:10:18:25:5
A15	Pruthvi Raj Chowk 2nd intersection		WB	9.8	24	270/80	42:10:23:25:2
A16	Pruthvi Raj Chowk 2nd intersection		NB	7	24	270/45	55:12:13:19:1
A17	Pruthvi Raj Chowk 2nd intersection		SB	7	24	270/45	54:10:16:18:2

Note: ** NB = Northbound; SB = Southbound; EB = Eastbound; WB = Westbound.

present study is formulating saturation flow model, traffic data at the selected intersections were collected during morning peak hour under clear weather condition. Based on observed data, the vehicles are classified into five categories: car (C), big car (BC), motorized two-wheeler (2W), three-wheeler (3W), and bus (HV). The key characteristics of the study locations along with the details of data collection and the vehicle composition during the study period are shown in Table 1.

Measurement of the Classified Vehicles during Saturated Green Time. For instance, $X = \{n_C, n_{BC}, n_{3W}, n_{2W}, n_{HV}, \frac{W}{w_b}\} \in \Re^6$ are the input variables where n_{bi} denotes classified vehicle counts; it indicates vehicle count for each vehicle category in saturated green time. n_{bi} is obtained by converting the total numbers of classified vehicles which crossed the stop line during the queue discharge time up to 20 s for each cycle from actual approach width (w) to standard lane width (w_b), due to the continuity of the model.

Under shared lane condition, it was observed that the rightmost portion of the effective road width was occupied by both through and right-turning vehicles. There were instances where no right-turners were present (approach 4 and 10), or a queue of right-turners had through vehicles standing behind them and vice-versa. Thus, remaining road width was used by through vehicles only as left-turners were being accommodated by the provision of left channelization. It needs mention that saturated green time for through and right turners ($g_{s-\text{th}}$ and $g_{s-\text{rt}}$) was measured separately for both straight and turning vehicles. Only straight-going vehicles were considered in the calculation of $g_{s-\text{th}}$. The saturated green time for through vehicles was measured when only through vehicles were passing. On the other hand, for the measurement of g_{s-th}, queue dissipation time for right-turning vehicles in saturated condition was noted down. During the data extraction process, it was ensured that no through vehicle was present in the queue of right-turning vehicles.

Consequently, both the saturation flow values (right and straight going) for effective approach width (w_{eff}) (i.e., consider for actual approach width for in this case), were converted to standard approach width (w_b) for better comparison. For this purpose, the effective approach widths are formulated from Equations 4 to 5 as

$$w_{eff-th} = w \times \frac{\textit{Total number of the straight through vehicles}}{\textit{Total number of the vehicles passed through the shared lane}} \quad (4)$$

$$w_{eff-rt} = w \times \frac{\textit{Total number of the right turning vehicles}}{\textit{Total number of the vehicles passed through the shared lane}} \quad (5)$$

where w_{eff-th} is the effective road width for straight through vehicles and w_{eff-rt} is the effective road width for right-turning vehicles. The assumption here is that the minimum effective approach width should be more than 1.8 m (maximum width of the passenger car).

Methodology for Estimation of Saturation Flow

The conventional saturation flow model mentioned above is easy to understand and hence, suitable for the on-field application. Present research work aims to investigate the effect of traffic composition, road width and right-turning vehicles on saturation flow at signalized intersections. Classified traffic volumes (n_{bi})(i.e., vehicle count for each vehicle category separately), the parameters that can describe total traffic volume as well as its composition characteristic within a saturated green period (g_s), are aptly chosen as design variables x in this case. In addition, f_w denotes the width ratio between actual approach width (w) and standard lane width (w_b), and is also considered to be a design variable in this case.

These design variables (x_{nm}) are only responsible for the changes in the response variables ($y(x)$), that is, saturation flow (or saturation flow for through and right-turning vehicles). This research proposes to model $y(x)$ by adopting the Kriging approximation (*29*).

Structure of Kriging Model. Before developing the Kriging-based saturation flow model, one should have a set of trial points/training points of design variables x and corresponding responses $y(x)$ given as

$$\boldsymbol{x} = \begin{bmatrix} x_{11} & \cdots & x_{1m} \\ \vdots & \ddots & \vdots \\ x_{n1} & \cdots & x_{nm} \end{bmatrix} \text{and } \boldsymbol{y}(\boldsymbol{x}) = \begin{bmatrix} y_1(\boldsymbol{x}) \\ \vdots \\ y_n(\boldsymbol{x}) \end{bmatrix} \quad (6)$$

where, m = number of design variable; and n = number of experiments. The unknown output (saturation flow in this case) using Kriging is represented as

$$y(\boldsymbol{x}) = y_0(\boldsymbol{x}) + \boldsymbol{Z}(\boldsymbol{x}) \quad (7)$$

where $y(x)$ represents the saturation flow, $y_0(x)$ is a trend function and $Z(x)$ represents the realization of a stochastic process with zero mean and nonzero covariance. In the model, the local deviation at an unknown point is expressed using stochastic processes. The sample points are interpolated with the Gaussian random function as the correlation function to estimate the trend of the stochastic processes. The $y_0(x)$ term is similar to a polynomial response surface and represented as

$$y_0(\boldsymbol{x}) = \boldsymbol{f}^T(\boldsymbol{x})\mathbf{b} \quad (8)$$

In the model, $f(x) = \{f_1(x), f_2(x), ...f_m(x)\}^T$ represents the functional vector and $b = \{b_1, b_2, ...b_m\}^T$ represents the regression coefficient vector where m is the number of the basis function of x. On contrary, $\boldsymbol{Z}(\boldsymbol{x})$ is the covariance of Gaussian stationary process represented as

$$\boldsymbol{Z}(\boldsymbol{x}) = \mathbf{Cov}(\boldsymbol{x}_i, \boldsymbol{x}_j) = \mathrm{Var}\mathbf{R}(\boldsymbol{x}_i, \boldsymbol{x}_j); \boldsymbol{i}, \boldsymbol{j} = 1, \ldots, \boldsymbol{m} \quad (9)$$

where, Var is the process variance and $\mathbf{R}(\boldsymbol{x}_i, \boldsymbol{x}_j)$ is the covariance between two points. The unknown coefficients **b** and process variance Var are estimated as

$$\mathbf{b} = \left(\mathbf{F}^{\mathrm{T}}\mathbf{R}^{-1}\mathbf{F}\right)^{-1}\mathbf{F}^{\mathrm{T}}\mathbf{R}^{-1}\mathbf{Y} \quad (10)$$

and

$$\mathrm{Var} = \frac{1}{\boldsymbol{n}}(\mathbf{Y} - \mathbf{Fb})^{-\mathrm{T}}\mathbf{R}^{-1}(\mathbf{Y} - \mathbf{Fb}) \quad (11)$$

where, **Y** represents observed responses at the data points. **F** and **R** in Equation 11 are computed as

$$\mathbf{F} = \begin{bmatrix} f_1(x_1) & \cdots & f_m(x_1) \\ \vdots & \ddots & \vdots \\ f_1(x_n) & \cdots & f_m(x_n) \end{bmatrix}, \quad \mathbf{R} = \begin{bmatrix} \mathrm{R}(x_1, x_1) & \cdots & \mathrm{R}(x_1, x_n) \\ \vdots & \ddots & \vdots \\ \mathrm{R}(x_n, x_1) & \cdots & \mathrm{R}(x_n, x_n) \end{bmatrix} \quad (12)$$

Algorithm 1. Algorithm for Kriging-Based Model

Initialize:

- Identify the input variables (say m).
  Select suitable correlation function. Collect data (influencing variables/input and saturation flow) from field (assume there are η_t training points)
- Formulate training matrix **F** (Equations 6 and 8), correlation matrix **R** (Equation 12) and Output/response vector **Y** (Equation 7).
- Compute unknown coefficient vector **b** and process variance *Var* from Equations 10 and 11.
- For a new set of samples, compute saturation flow by using Equation 15.

where, $\mathrm{R}(\mathrm{x_i}, \mathrm{x_j})$ is a correlation function for n number of data set. There exist several correlation functions to be utilized within the Kriging. In most of the cases, the correlation function is stationary (only depends on the distance between the training samples and not on their location) and hence, can be decomposed as

$$\mathrm{R}(x_i, x_j) = \prod_{n=1}^{m} R_n\left(\theta_n, x_i^n - x_j^n\right) \tag{13}$$

where, θ_n, $n = 1,\ldots,m$ are the unknown correlation parameters used to fit the model for n design variables and $(x_i^n - x_j^n)$ represents the distance between two points. One of the important features of the Kriging approach is to obtain the hyperparameters, which control the correlation functions. In this study, the maximum likelihood (ML) estimator is used to determine θ_n. Ultimately, based on observed field data, the unknown regression coefficient vector (**b**) and process variance (Var) are computed by using Equations 10 and 11.

In this work, the Gaussian correlation function has been used due to its superior performance (*23*).

$$\mathrm{R} = \exp\left[-\sum_{n=1} \theta_n \left|x_i^n - x_j^n\right|^2\right] \tag{14}$$

Prediction of the saturation flow due to given inputs. Once the unknown coefficient vector **b** and process variance Var are determined, the next step of the methodology deals with the response at an unknown point (new design variable x) which can be predicted as

$$y(\boldsymbol{x}) = \boldsymbol{f}^{\boldsymbol{T}}(\boldsymbol{x})\mathbf{b} + \mathbf{r}^{\mathrm{T}}(\mathbf{x})\mathbf{R}^{-1}(\mathbf{Y} - \mathbf{Fb}) \tag{15}$$

where, unknown regression coefficient vector b in Equation 14, has been computed by using Equation 10; $\boldsymbol{f}^{\boldsymbol{T}}(\boldsymbol{x})$ is estimated by considering the new set of values (x) into the functional vector; new correlation function for the new set of points as given as

$$\mathbf{r}^{\mathrm{T}}(\mathbf{x}) = \{R(x, x_1), R(x, x_2), \ldots..R(x, x_n)\} \tag{16}$$

An algorithm showing the step-by-step procedure of Kriging-based saturation flow model is shown in Algorithm 1.

Adjustment Factor Due to Shared Right-Turning and Through Vehicles

It is noted that there was a reduction in saturation flow when right-turning vehicles (for left-hand drive) were present. The Canadian Capacity Guide (*18*) established the saturation flow adjustment factors by considering the vehicles composition and turning vehicles proportion at a shared lane. It was observed that saturation flow for right-turning movement and saturation flow for through movement are significantly different in heterogeneous traffic condition. Based on this observation, an adjustment factor considering both shared right-turning and through vehicles is proposed in this research. This adjustment factor at an intersection can be represented as

$$F_{\mathrm{rt}} = \frac{(S_{\mathrm{rt}} + S_{\mathrm{th}})}{S'_{\mathrm{eth}}} \tag{17}$$

where

F_{rt} = adjustment factor for shared right-turning and through vehicles;

S_{rt} = right-turning saturation flow in shared lane in terms of PCU/hr/lane;

S_{th} = through saturation flow in shared lane in PCU/hr/lane; and

S'_{eth} = equivalent through saturation flow in shared lane (PCU/hr/lane), which is computed from Equation 18 as

$$S'_{\mathrm{eth}} = k_{\mathrm{rt}} S_{\mathrm{rt}} + S_{\mathrm{th}} \tag{18}$$

where, k_{rt} is the right-turning movement factor which is the ratio of the through saturation flow and right-turning saturation flow in the shared lane as expressed in Equation 19 as

$$k_{\mathrm{rt}} = \frac{S_{\mathrm{th}}}{S_{\mathrm{rt}}} \tag{19}$$

Algorithm 2. Evaluation of Saturation Flow Adjustment Factor for Shared Right-Turn and through Movements

Initialize:
1) Select site and collect data
2) Determine the saturation flow for the right-turn and through movements separately by Kriging model (Algorithm 1)
3) Utilizing Equation 19, compute the right-turning movement factor, k_{rt}
4) Based on observed field data, determine the equivalent through saturation flow in shared lane, S'_{eth} by using Equation 18
5) Compute saturation flow adjustment factor for shared right-turn and through movements (Equation 17).

An algorithm describing the procedure outlined above is shown in Algorithm 2.

Results and Discussion

Saturation flow model based on Kriging has been formulated by using classified (category wise) vehicle count $(n_{(\cdot)})$ and approach width ratio $\left(\frac{W}{w_b}\right)$ as the influential parameters and hence, $X = \{n_C, n_{BC}, n_{3W}, n_{2W}, n_{HV}, \frac{W}{w_b}\} \in \Re^6$, where the subscript C, BC, 3W, 2W and HV represents car, big car, three-wheeler, two-wheeler and heavy vehicle respectively. The saturation flow for a whole approach width (s) (in terms of PCU/hr) was taken as output response. As already stated, Gaussian covariance function has been used for formulating the Kriging-based saturation flow model. To assess whether Kriging model provides the best fit, it is necessary to check its prediction accuracy based on the external dataset. Additionally, to illustrate the superior performance of the proposed approach, results obtained using the regression method have also been reported.

Figure 1 shows the scatter plot of saturation flow obtained using the proposed Kriging model. As expected, the Kriging-based model predicts superior results as compared to the existing regression-based saturation flow model. To be specific, while the regression-based model yields a coefficient of determination of 0.6672, proposed model based on Kriging yields coefficient of determination of 0.8657. Hence, one may confer that the proposed Kriging model is better as far as accuracy is concerned.

Table 2 shows the detailed comparison of the mean saturation flow values for the whole approach width obtained by the conventional method, regression method and Kriging method for validation purpose. It was observed that higher standard deviation of the saturation flow was obtained from the conventional method and existing regression method as compared with the standard deviation of the saturation flow by Kriging method. The corresponding mean absolute percentage error (MAPE) values with respect to the conventional method have also been computed. Additionally, mean saturation flow per 3.5 m (Indian standard road width) per hour has been calculated. In both the cases, it can be observed that the saturation flow values produced by the proposed Kriging method are closer to the observed saturation flows using conventional method (i.e., the one directly computed based on field data). In addition, the

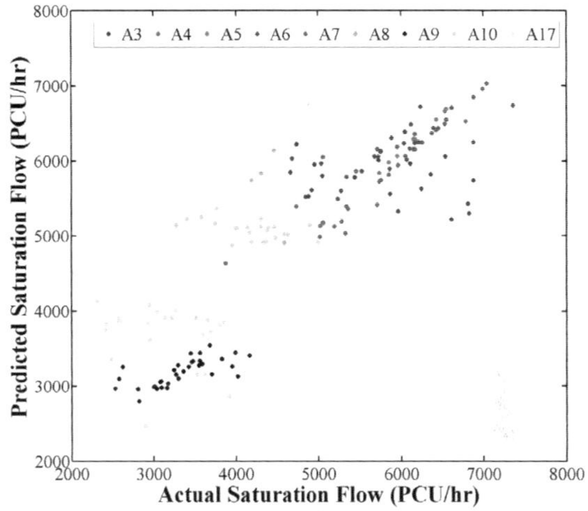

(a)

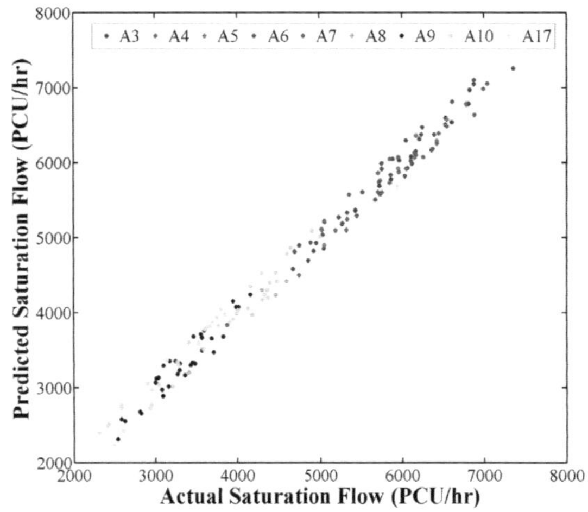

(b)

Table 2. Comparison of Saturation Flow from Different Methods

Approach ID	Mean saturation flow of the approach in PCU/h					Mean saturation flow in PCU/h/lane$		
	From conventional method	From regression method		From Kriging method		From conventional method	From regression method	From Kriging method
	Saturation flow/SD	Saturation flow/SD	MAPE, %	Saturation flow/SD	MAPE, %			
A1	4298.20/487	4146.05/451	7.21	4176.02/211	1.23	1842.14	1813.9	1827.01
A2	4851.7/356	4637.48/297	1.8	4919.69/ 196	2.75	1821.47	1764.26	1871.62
A3	6070.99/601	5517.53/448	2.31	5851.97/189	0.25	2004.21	1893.27	2008.03
A4	5477.34/559	5752.79/601	5.03	5433.14/437	1.01	1976.36	2075.75	1960.41
A5	6312.88/719	5739.69/312	2.04	6273.75/289	1.48	2056.91	1877.47	2052.16
A6	6065.29/881	6089.27/659	3.35	5963.11/277	1.83	1921.22	1955.27	1914.76
A7	5634.9/431	5067.75/233	5.14	5026.27/201	4.28	1874.37	1970.79	1954.66
A8	4894.2/576.01	4938.66/522	3.29	4997.8/231	3.65	1875.18	1920.59	1943.59
A9	3055.2/447	3253.59/397	4.24	3292.58/176	2.15	1980.77	1897.93	1920.67
A10	5142.84/354	5081.66/301	2.98	4919.63/219	3.57	1999.99	1976.2	1913.19
A17	4043.32/495	3788.9/227	5.56	3907.13/210	3.31	1953.03	1894.45	1908.33

Note: $Standard approach road width = 3.5 m for urban road as per IRC-86 (*30*); SD = standard deviation.

corresponding MAPE values of the saturation flow values based on Kriging method are quite low (less than 5%) compared with the existing regression method.

Influence of Traffic Composition and Approach Width on Saturation Flow

After validation of the proposed model, these models were utilized to predict the saturation flow at varying approach width and traffic composition. The values of influencing variables were varied in such a way that they could reflect the impacts of actual approach width and traffic composition on saturation flow at the signalized intersections. Therefore, the effect of traffic composition on saturation flow per standard road width of an intersection has been investigated in this section. Figure 2 shows the variation of saturation flow with a change in vehicle composition corresponding to 3.5 m standard approach road width (as per Indian standard) respectively. In this process, the proportion of cars in the traffic stream has been varied from 20% to 50%. On the contrary, the proportion of big cars, the proportion of 3Ws, the proportion of 2Ws and proportion of HVs has been varied from 5% to 20%, 5% to 15%, 20% to 50%, and 3% to 15% respectively. It is observed that the proportion of the cars, 3Ws and 2Ws in the traffic stream, are the most sensitive one affecting the saturation flow. In addition, due to the presence of both slow- and fast-moving vehicles in the traffic stream, extremely poor lane discipline is usually observed in a highly heterogeneous traffic situation at signalized intersections located in urban areas. To be specific, an increase in the proportion of 3Ws and 2Ws decreases the saturation flow per standard width. Motorized 2Ws and 3Ws can move through small gaps and zigzag direction with high acceleration during green phase due to their small size and better maneuverability. As a result of that, movements of other vehicle categories get interrupted leading to a lower saturation flow. However, an increase in the proportion of cars, BC and HV result in an increase of the saturation flow at the intersection. HV has poor dynamic characteristics compared with other vehicles, and therefore saturation flow rises at flatter rate with an increase of HVs.

Approach road width was varied from 3.5 to 11 m in steps of 2.5 m, and correspondingly, the saturation flow was estimated using the model. Figure 2*d* shows saturation flow increases with the increase in the approach road. However, the relationship between approach width and saturation flow is not exactly linear. It indicates the cumulative effect of the lane indiscipline behavior. At non-lane discipline traffic condition, all vehicles try to utilize every gap available to them. It results in more efficient movement of vehicles, and thus, more discharge per unit width of the approach takes place as the approach width increases, and freedom to individual vehicles becomes more.

Influence of Right-Turning Movement on Saturation Flow

Saturation flow values per standard approach road width with a single lane for both through (s_{th}) and right turn (s_{rt}) movements are estimated by using conventional method (Equation 2) as well as Kriging technique (Equation 15) in terms of PCU/h/lane. In the case of saturation flow for through vehicles, classified traffic volumes ($n_{i\text{-}th}$) have been measured by considering

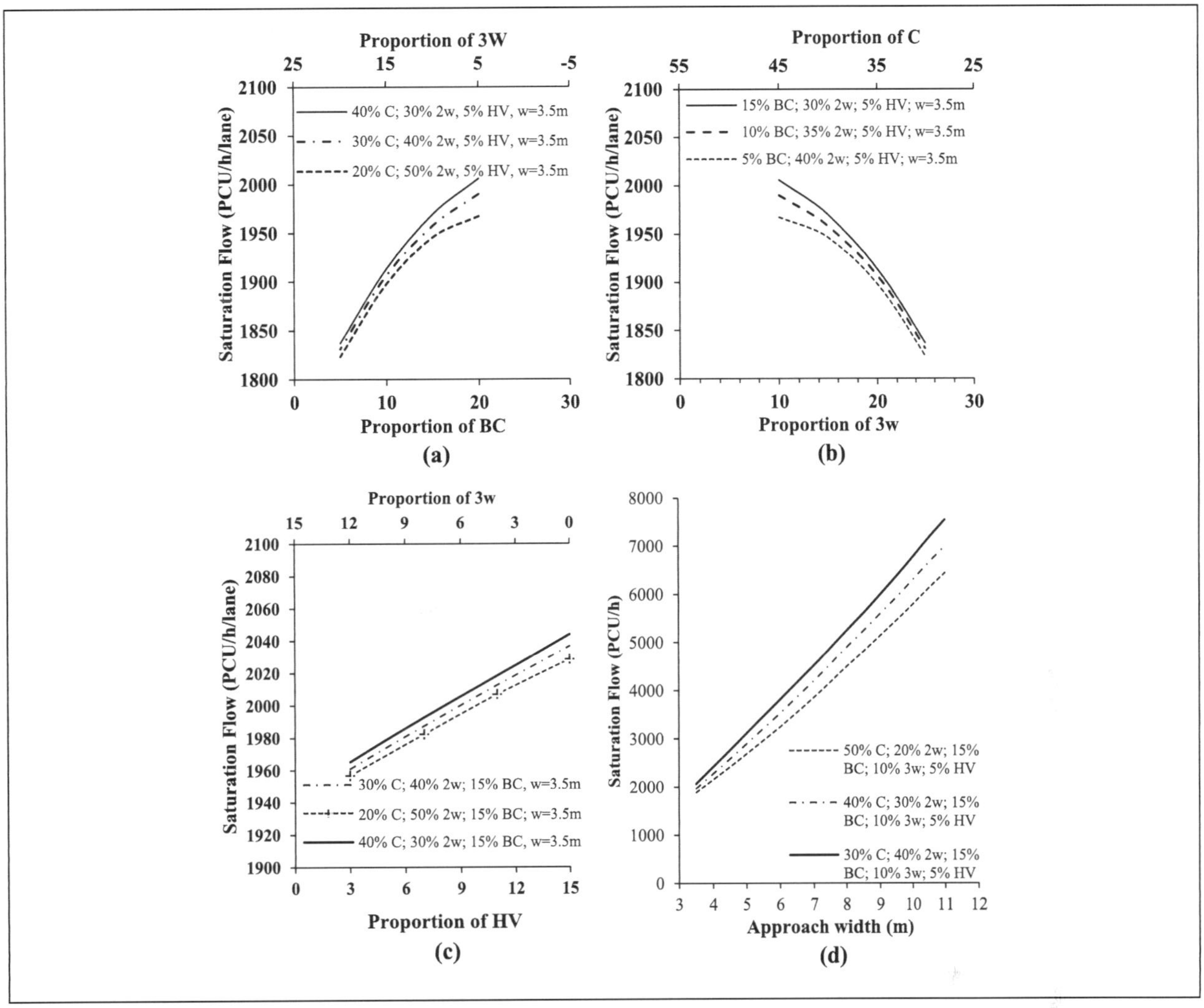

Figure 2. Influence of traffic composition and approach road width on the saturation flows.

saturated green period for respective movements ($g_{s\text{-th}}$). Similarly, classified traffic volumes ($n_{i\text{-}rt}$) for right-turning vehicles were measured separately as mentioned earlier. Therefore, the classified vehicles count $X = \{n_{C-\mathrm{th}}, n_{BC-\mathrm{th}}, n_{3W-\mathrm{th}}, n_{2W-\mathrm{th}}, n_{HV-\mathrm{th}}, n_{C-\mathrm{rt}}, n_{BC-\mathrm{rt}}, n_{3W-\mathrm{rt}}, n_{2W-\mathrm{rt}}, n_{HV-\mathrm{rt}}\} \in \Re^{10}$ is considered to be the number of design variables of the movement-wise saturation flow prediction model. The saturation flows for movement wise s_i(i = rt, th) (in terms of PCU/hr/lane) were taken as output responses. These corresponding effective approach width, based on the proportion of right-turning vehicles and saturation flow for the approaches, are given below in Table 3.

It is observed that right-turning movements have an influence on the saturation flow for an approach. Consequently, right-turning movements were analyzed separately at a signalized intersection and the adjustment factor is computed for shared right-turn and through movements (Table 3). It is multiplied by the saturation flow value to address the right-turning movement's effects. Right-turning movements were restricted in a few approaches (A4 and A10). In another case (A9), effective approach widths were found to be less than 1.8 m (by using Equation 5) due to the presence of a lower proportion of right-turning vehicles. Hence, the adjustment factors were not calculated for these approaches.

Conclusion

This study presented a model for predicting saturation flow of an intersection based on categorical vehicle count and approach road width. The primary idea is to utilize Kriging, an alternative to the conventional regression technique, for developing a saturation flow model.

Table 3. Estimation of Saturation Flow Adjustment Factor for Shared Right-Turn and Through Movements

	Vehicles proportion (%)	Effective approach width (m)	Saturation flow (PCU/h/lane) by conventional method			Saturation flow (PCU/h/lane) by Kriging method		
Approach ID	th/rt	th/rt[##]	s_{th}	s_{rt}	F_{rt}	s_{th}	s_{rt}	F_{rt}
A1	0.69/0.31	5.5/2.5	1880.46	1799.96	**0.979**	1830.90	1748.49	**0.977**
A2	0.511/0.48	4.74/4.46	1845.75	1669.17	**0.986**	1854.26	1635.21	**0.941**
A3	0.75/0.25	7.7/2.5	2083.18	1857.21	**0.962**	1953.27	1778.23	**0.955**
A4	1/0	9.7/0	1976.36	NA	**1.000**	2075.75	NA	**1.000**
A5	0.81/0.19	8.57/2.13	2064.96	1932.65	**0.996**	1897.77	1859.54	**0.990**
A6	0.82/0.18	8.9/2	1947.57	1917.125	**0.986**	1975.67	1916.71	**0.985**
A7	0.69/0.31	6.2/2.8	2191.34	1932.65	**0.855**	2270.79	1796.67	**0.896**
A8	0.78/0.22	7/2.0	1903.3	1912.65	**0.985**	1927.26	1916.71	**0.997**
A9	0.88/0.12	5.3/0.7	1980.79	NA	**1.000**	1897.93	NA	**1.000**
A10	1/0	9/0	1999.99	NA	**1.000**	1976.2	NA	**1.000**
A17	0.7/0.3	4.9/2.1	2021.65	1861.23	**0.966**	1981.11	1872.46	**0.973**

Note: [##]th = through movement; rt = right-turning movement.

Kriging performs a bi-level approximation: first on a global scale using a trend function and second on a local scale by using covariance function. For this reason, Kriging yields much superior results compared with conventional regression-based methods. Moreover, to account for the effect of right-turning vehicles, a new right-turning factor has also been proposed. It is demonstrated that the proposed model yields superior results compared with conventional regression techniques. Additionally, the proposed model also serves as a tool to minimize the standard deviation in saturation flow. The major findings of this study are summarized as follows:

- The approach width has a strong influence on saturation flow at the signalized intersection. However, the saturation flow does not linearly increase as approach width increases.
- With the increase of approach road width, freedom to individual vehicles becomes more due to non-lane discipline traffic condition. It is the outcome of the cumulative effect of the lane indiscipline behavior.
- For the same approach road condition, the proportions of 3Ws and 2Ws most influence the predicted saturation flow. The increase in the percentage of small vehicles results in less space occupancy on the road. For the better maneuverability, small vehicles move through small gaps in a zigzag manner with high acceleration. As a result, they interrupt the movements of other vehicles. Thus, it leads to a reduction in the saturation flow at the signalized intersection. On the other hand, an increase in the proportion of HVs increases the saturation flow at the same approach width. However, the influence of HVs is marginal due to their poor operational characteristics.
- Further, it is seen from the present study that the right-turning movements have an influence on the saturation flow for an approach. Unlike available adjustment factors, the proposed adjustment factor is formulated by considering classified vehicle count for both right-turning and through movements separately to be the significant parameters. Therefore, it can accurately evaluate the right-turning effect under heterogeneous traffic condition.

The whole idea of this work is to use Kriging-based model to tackle the effect of traffic composition and road width on saturation flow and to determine adjustment factors for all other non-ideal cases. the focus of the research is on developing saturation flow model with traffic composition, road width and presence of right-turning vehicles as influential factors. Therefore, the sites were selected in accordance with the objective of the research only (i.e., other factors are not present). However, in situations where all the factors are present, the same approach can easily be extended by using Kriging to handle traffic composition and road width and adjustment factors for all the other cases. Obviously, suitable adjustment factors need to be computed for the same, which is beyond the scope of the present paper, but can certainly be taken up in future research

Author Contributions

The authors confirm contribution to the paper as follows: study conception and design: Sabyasachi Biswas (SB), Souvik Chakraborty (SCh), Indrajit Ghosh (IG) and Satish Chandra (SC); data collection: SB; analysis and interpretation of results: SB and SCh; draft manuscript preparation: SB and SCh. All authors reviewed the results and approved the final version of the manuscript.

References

1. *Highway Capacity Manual 2010*. Transportation Research Board of the National Academies, Washington, D.C., 2010.
2. Greenshields, B. D., D. Shapiro, and E. L. Erickson. *Traffic Performance at Urban Intersections, Bureau of Highway Traffic*. Technical Report No. 1. Yale University, New Haven, 1947.
3. Tarko, A. P., and M. Tracz. Uncertainty in Saturation Flow Predictions. *Transportation Research Circular E-C018*, 2000, pp. 310–321.
4. *Highway Capacity Manual*. TRB, National Research Council, Washington, D.C., 2000.
5. Scraggs, D. A. *The Passenger Car Equivalent of a Heavy Vehicle in Single Lane Flow at Traffic Signals*. Note No. LN 573/DAS, U.K. London, U.K., 1964.
6. Niittymaki, J., and M. Pursula. Saturation Flows at Signal-Group-Controlled Traffic Signal. *Transportation Research Record: Journal of the Transportation Research Board*, 1997. 1572: 24–32.
7. Miller, A. J. *Australian Road Research Board Bulletin No. 3*. Melbourne, 1968.
8. Webster, F. V., and B. M. Cobbe. *Traffic Signal, Ministry of Transport, Road Research*. Technical Paper No. 56, 1966.
9. Li, H., and P. Prevedouros. Detailed Observations of Saturation Headways and Start-Up Lost Times. *Transportation Research Record: Journal of the Transportation Research Board*, 2006. 1802: 44–53.
10. Indian Roads Congress. Guidelines for the Design of at Grade Intersections in Rural and Urban Areas. *IRC Code of Practice*, Vol. SP41, 1994.
11. Bhattacharya, P. G., and A. K. Bhattacharya. Observation and Analysis of Saturation Flow Through Signalised Intersections in Calcutta. *Indian Highways*. Indian Roads Congress, No. April, 1982, pp. 11–33.
12. Rahman, M., S. Nur-Ud-Deen, and T. Hassan. Comparison of Saturation Flow Rate at Signalized Intersections in Yokohama and Dhaka. *Proceeding of the Eastern Asia Society for Transportation Studies*, Vol. 5, No. 1986, 2005, pp. 959–966.
13. Hadiuzzaman, M., M. Rahman, and M. A. Karim. Saturation Flow Model at Signalized Intersection for Non-Lane Based Traffic. *Canadian Journal of Transportation*, Vol. 2, No. 1, 2008.
14. Transport and Road Research Laboratory. *A Method of Measuring Saturation Flow at Traffic Signals*. Transport and Road Research Laboratory, London, 1963.
15. Turner, J., and G. Harahap. Simplified Saturation Flow Data Collection Methods. *Proc., CODATU VI Conference on the Development and Planning of Urban Transport*, Tunis, 1993, pp. 1–15.
16. Holroyd, E. M. Effect of Motorcycles and Pedal Cycles on Saturation Flow at Traffic Signals. *Roads Road Construction*, Vol. 41, 1963, pp. 315–616.
17. Branston, D. M., and P. Gipps. Some Experience with a Multiple Linear Regression Method of Estimating Parameters of the Traffic Signal Departure Process. *Transportation Research*, Vol. 15, No. 6, 1981, pp. 445–458.
18. Teply, S., D. I. Allingham, D. B. Richardson, B. W. Stephenson, and J. W. Gough. *Canadian Capacity Guide for Signalized Intersections*, 3rd ed. Institute of Transportation Engineers, 2008.
19. Hansson, A., and T. Bergh. *A New Swedish Capacity Manual CAPCAL 2*, 1988.
20. Chang, G.-L., C.-Y. Chen, and C. Perez. Hybrid Model for Estimating Permitted Left-Turn Saturation Flow Rate. *Transportation Research Record: Journal of the Transportation Research Board*, 1996. 1566: 54–63.
21. Chen, P., Q. Hongsheng, and S. Jian. Investigation of Saturation Flow on Shared Right-Turn Lane at Signalized Intersections. *Transportation Research Record: Transportation Research Board*, 2014. 2461: 66–75.
22. Giannopoulos, G. A. Saturation Flow and Capacity of Shared Lanes. *Transportation Research Record: Journal of the Transportation Research Board*, 1996. 1555: 50–58.
23. Kaymaz, I. Application of Kriging Method to Structural Reliability Problems. *Structural Safety*, Vol. 27, No. 2, 2005, pp. 133–151.
24. Dubourg, V., B. Sudret, and J.-M. Bourinet. Reliability-Based Design Optimization Using Kriging Surrogates and Subset Simulation. *Structural and Multidisciplinary Optimization*, Vol. 44, No. 5, 2011, pp. 673–690.
25. Branston, D. M. Some Factors Affecting the Capacity of Signalized Intersections. *Traffic Engineering and Control*, Vol. 20, No. 8/9, 1979, pp. 390–396.
26. Kimber, R. M., M. McDonald, and N. B. Hounsell. Passenger Car Units in Saturation Flows: Concept, Definition, Derivation. *Transportation Research Part B: Methodological*, Vol. 19, No. I, 1985, pp. 39–61. https://doi.org/10.1016/0191-2615(85)90028-1
27. Biswas, S., and I. Ghosh. Estimation of the Passenger Car Units and Stream Equivalency Factor at Signalized Intersections Under Heterogeneous Traffic Condition. Presented at 96th Annual Meeting of the Transportation Research Board, Washington, D.C., 2017.
28. Biswas, S., I. Ghosh, and S. Chandra. Effect of Countdown Timers on Traffic Characteristics and Driver Behaviors at Signalized Intersections. Presented at 95th Annual Meeting of the Transportation Research Board, Washington, D.C., 2016.
29. Krige, D. G. A Statistical Approach to Some Basic Mine Valuation Problems on the Witwatersrand. *Journal of the Chemical Metallurgical and Mining Society of South Africa*, Vol. 52, No. 6, 1951, pp. 119–139.
30. Indian Roads Congress. Geometric Design Standards for Urban Roads in Plains. *IRC Code of Practice*, Vol. 86, 1983, p. 38.

The Standing Committee on Highway Capacity and Quality of Service (AHB40) peer-reviewed this paper (18-03975).

Article

TRR
JOURNAL OF THE TRANSPORTATION RESEARCH BOARD

Transportation Research Record
2018, Vol. 2672(15) 66–74

DOI: 10.1177/0361198118774743
journals.sagepub.com/home/trr
SAGE

Traffic Operations on Rural Two-Lane Highways: A Review on Performance Measures and Indicators

Ahmed Al-Kaisy[1], Amirhossein Jafari[1], Scott Washburn[2], Tapio Lutinnen[3], and Richard Dowling[4]

Abstract

Two- lane highways constitute a large proportion of the highway system in the United States, particularly in rural areas. Performance evaluation on those highways is the basis for planning, upgrade, and improvement programs. Such evaluations are conducted using performance measures that should ideally be correlated to performance determinants on those highways. Unlike other highway facilities, two-lane highway performance is notably affected by the platooning phenomenon, which is believed to be a function of traffic level and passing opportunities. Over the last couple of decades, several studies have addressed performance measures on two-lane highways, and are mainly driven by the reported limitations of the current measures used by the *Highway Capacity Manual* (HCM). This review aims to summarize the performance measures that have been published in the literature or used in practice both in the United States and abroad. The review starts with an historical overview of the HCM performance measures since its inception in 1950. It then discusses the most important criteria for measures to be more effective in describing performance, before presenting the review results for measures that have been proposed in literature or reported as being used in practice. Finally, a subjective assessment of all performance measures against the set of criteria outlined in the article is presented. The review presented in this paper provides information that is valuable for practitioners and researchers in understating the alternative measures for assessing performance on two-lane highways, and the limitations and merits associated with those measures.

Two-lane, two-way highways constitute a large proportion of the highway system, particularly in rural areas. The unique characteristic that separates two-lane highways from other uninterrupted flow facilities is that passing occurs on the opposing lane of traffic. There should be an adequate gap and sufficient sight distance in the opposite direction of traffic to enable a passing maneuver to occur. Passing opportunities decrease as flow rate and geometric restrictions increase. Lack of passing opportunities typically results in the formation of platoons with trailing vehicles subject to additional delay.

The level of service (LOS) is a qualitative measure used by the *Highway Capacity Manual* (HCM) to show the quality of traffic flow on a highway facility. Currently, the HCM uses a scale from A to F to show the LOS, with A defined as the best and F defined as the worst condition for traffic flow. Performance measures are used to assess operations and LOS. On two-lane highways, they are intended to reflect the level of freedom for drivers to select their desired speeds and pass slow-moving vehicles. Measuring performance is important for operational analysis and for upgrade and improvement projects. A low LOS on a section of a two-lane highway may be improved by adding a passing lane or expanding the route into a continuous three lane section or a multi-lane highway.

Operational analysis on two-lane highways in the United States and Canada (and some other countries outside North America) is usually conducted using the procedures outlined in the HCM (*1*). In the most recent edition of the HCM (*1*), average travel speed (ATS) and percent time spent following (PTSF) are used for evaluating performance on Class I two-lane highways. Class I

[1]Department of Civil Engineering, Montana State University, Bozeman, MT
[2]Department of Civil and Coastal Engineering, University of Florida, Gainesville, FL
[3]Department of Civil and Environmental Engineering, Aalto University, Finland
[4]Kittelson & Associates, Inc., Oakland, CA

Corresponding Author:
Address correspondence to Ahmed Al-Kaisy: aalkaisy@ce.montana.edu

Table 1. LOS Criteria for Two-Lane Highways (*1*)

LOS	Class I highways		Class II highways	Class III highways
	ATS (mph)	PTSF (%)	PTSF (%)	PFFS (%)
A	>55	≤35	≤40	>91.7
B	>50–55	>35–50	>40–55	>83.3–91.7
C	>45–50	>50–65	>55–70	>75.0–83.3
D	>40–45	>65–80	>70–85	>66.7–75.0
E	≤40	>80	>85	≤66.7

highways include major intercity routes, daily commuter routes, and major links in state or national highway networks. ATS refers to the average speed of all vehicles moving through the highway measured at a point. PTSF is defined as "the average percentage of travel time that vehicles must travel in platoons behind slower vehicles because of an inability to pass" (*1*). PTSF is also used for Class II two-lane highways, which comprise highways that serve as access routes to Class I facilities as well as scenic and recreational routes. On those highways, motorists do not necessarily expect to travel at high speeds. Due to difficulties in measuring the PTSF in the field, the HCM recommends using a surrogate measure, calculated as the percentage of vehicles following other vehicles with headways smaller than 3 s. The third measure, percent free-flow speed (PFFS), is used for evaluating performance on Class III highways following recommendations from Washburn et al. (*2*). It is the ratio of ATS to free-flow speed. Class III highways are two-lane highways serving moderately developed areas. They may include portions of Class I and Class II highways that pass through small towns or developed recreational areas. Table 1 shows the LOS criteria for two-lane highways in the 2016 HCM (*1*).

Several studies reported limitations in the current performance measures used by the HCM (*3–14*). The main performance measure, PTSF, is limited in that it is difficult to measure in the field, is not compatible with the service measures of other facilities, does not describe the extent of congestion, and is not very useful in other analyses (e.g., economic, environmental, etc.). Nor is PTSF a good performance measure for indicating if improvements should be made to a highway that has low volumes but a high percentage of heavy vehicles and few passing opportunities. ATS, on the other hand, is not very informative about the efficiency of the highway. As the analysis section of a two-lane highway facility is usually several miles long, there could be many changing conditions, such as posted speed limit and roadway alignment that affect ATS, yet they are not related to varying traffic conditions. The PFFS, an indicator of speed reduction due to traffic, can easily be measured in the field; however, it is not very sensitive to flow level for the uncongested traffic regime. Therefore, adding a passing lane was not found to tangibly affect this performance measure, as reported by Al-Kaisy and Freedman (*3*).

Several alternative performance and service measures for two-lane highways have been proposed in the literature. Most of the studies that proposed new performance measures were driven by the obvious limitations of the HCM performance measures used. This paper provides a review of alternative performance measures that have been proposed in the literature or reported as part of today's practice. All these measures are intended for estimating performance in a directional traffic stream, i.e., in one direction of a two-lane, two-way highway.

HCM Performance Measures on Two-Lane Highways: Historical Overview

The HCM 1950 (*15*) was the first document introducing an analytical method for capacity analysis on two-lane highways. Practical capacity, which corresponds to the capacity under reasonable driving conditions, was used in this manual to assess the operating conditions on two-lane highways. Operating speed was the indicator for measuring the practical capacity. The concept of LOS was presented in the second edition of the HCM in 1965 (*16*). Operating speed and volume to capacity (v/c) ratio were used to estimate the LOS on two-lane highways. In 1985, the HCM introduced percent time delay (PTD) for measuring performance on two-lane highways in addition to ATS (*17*). It refers to the percentage of time that vehicles are delayed and trapped in platoons due to an inability to pass slower vehicles. Delay refers to situations in which drivers travel at speeds lower than their desired speeds. A headway threshold of 5 s was used to estimate the PTD in the field. The fourth edition of the HCM introduced PTSF which is a similar measure to PTD and refers to the percent of time vehicles are following other vehicles on two-lane highways (*18*). The percentage of vehicles with headways less than 3 s was used as a surrogate measure for PTSF in this edition of the HCM. The following edition of the HCM was released in 2010 (*19*) and introduced a new classification for two-lane highways: classes I, II, and III. In addition to PTSF and ATS, PFFS was the new measure introduced in this edition for operational analysis of Class III two-lane highways. The most recent edition of the HCM was released in 2016 (*1*). The analytical methodology including performance measures in this edition is similar to that in the 2010 edition.

What Consitutes a Good Perfrormance Measure on Two-Lane Highways?

Ideally, performance measures used for traffic operations and capacity analysis on any highway facility should meet the following criteria (*5*):

1. The performance measure should reflect the perception of road users on the quality of traffic flow. It is a common understanding that platooning and lack of passing opportunities (and the associated delay) primarily affect the motorists' perception of the quality of service on two-lane highways.
2. The performance measure should be easy to measure and estimate using field data. Specifically, it is expected that the prospective performance measure can be estimated in the field using conventional data collection methods used by highway agencies and the professional community. Although some spatial performance measures are more difficult to observe in the field (e.g., density, ATS, etc.), it has been an accepted practice to estimate those measures using conventional data collection techniques at point locations.
3. The measure should correlate to traffic and roadway conditions in a meaningful way. On two-lane highways, the prospective performance measure should closely correlate to the platooning phenomenon (and passing opportunities) as well as to traffic level in a logical and meaningful way.
4. It is recommended that the prospective measure be compatible with performance measures used on other facilities. This criterion may be hard to satisfy as it is expected that different aspects of traffic operations are perceived as most important by drivers on different highway facilities. For example, although platooning on two-lane highways is a major determinant of the quality of operation, it is not a major factor in determining the quality of operations on other facilities.
5. The performance measure should be able to describe both uncongested and congested conditions. Although this requirement is applicable to all highway facilities as per the definition of the LOS scheme, it has been perceived to have less significance on two-lane highways, as these facilities are rarely congested and are usually upgraded to four lanes at traffic levels well below capacity.
6. The measure should be useful in analyses concerning traffic safety, transport economics, and environmental impacts. Increasingly, the capacity analysis procedures have been used in supporting the aforementioned analyses. To be of use in safety analyses, the prospective measure should correlate well to traffic exposure. On the other hand, for economic and environmental impact analyses, the prospective measure should be useful in estimating delay and its associated fuel consumption and tailpipe emissions.

The six criteria above consider the common operational objectives of most highway agencies, namely mobility, productivity, safety, reliability, and environmental impacts.

Review Results

This study identified many performance measures for two-lane highways that are proposed in the literature or reported as being used in practice. In this section, performance measures used by the HCM are presented first, followed by those that have seen application in other countries, and finally, alternative measures that have been proposed in the literature.

HCM Performance Measures

The HCM has been the principal reference for performing capacity analyses in the United States and Canada since its inception in 1950. Therefore, the performance measures used by the current edition of the HCM largely represent the current practice in the United States and Canada. As stated earlier in this article, three performance measures are used by the current HCM procedures for two-lane highways: PTSF, ATS, and ATS as a percentage of free-flow speed (FFS).

Percent Time Spent Following. This performance measure has been in use since the fourth edition of the HCM that was published in 2000 (*18*). Although the PTSF concept correlates well with the platooning phenomenon, the major determinant of performance on two-lane highways, it is almost impractical to measure in the field. Given this limitation, the HCM suggests using the percentage of headways less than 3 s as a good estimate for PTSF using field data. However, though PTSF by definition is a spatial measure, that is, measured over a stretch of two-lane highways, the percentage of headways less than 3 s can only be measured at a point location.

Average Travel Speed. The vast majority of two-lane highways can be thought of as "uninterrupted flow facilities," thus enjoying relatively higher travel speeds. This is particularly true for the HCM Class I highways, which represent important arterials and major collectors in rural areas. On these highways, ATS has long been used by the HCM as a performance measure with the premise that average speed is affected by traffic level and, thus, the amount of platooning due to limited passing opportunities. However, two-lane highways involve most highway classifications, have a wide range of geometric standards, and consequently, a wide range of operating speeds. Therefore, using average speed alone may not provide enough information about the level of traffic performance (in the absence of a reference point) to make performance comparisons across sites practical. ATS is also used as a performance measure for two-lane highways in Brazil (*20*). A recent survey of practice found that ATS is perceived to be the most relevant traffic flow aspect to two-lane highway operations by HCM users (*21*).

ATS as a Percentage of FFS. This performance measure has been in use since the HCM fifth edition, published in 2010, for use on Class III two-lane highways (*19*). Using ATS as a percentage of FFS was viewed as a good indicator of the amount of speed reduction due to traffic and the amount of vehicular interaction in the traffic stream. However, evaluations using field data in the United States, Egypt, and Spain (*6*, *8*, *9*) showed that the speed measures did not exhibit good correlations with platooning variables compared with other performance measures investigated in these studies. Furthermore, Luttinen (*5*) reported on an older study by Kiljunen and Summala in 1996 that proposed the use of ATS/FFS as a performance measure on Finnish two-lane highways.

Performance Measures Used in Other Countries

In this section, the major performance measures used in other countries are presented and discussed. HCM performance measures that found applications in practice outside the US are not covered in this discussion.

In their article on the German experience, Brilon and Weiser (*22*) reported that the current German Capacity Handbook (*23*) utilizes density as the primary service measure for two-lane highways. The rationale for using density as a performance measure on two-lane highways in Germany is that efficiency is given preference over user experience (perception) of the quality of service (*22*). Furthermore, this performance measure is compatible with other facility types, mainly freeways and multi-lane highways, when those highway types are analyzed as part of a larger system. For this purpose, density is calculated as the ratio of traffic volume and the ATS of just passenger cars (i.e., ATS_{pc}). Average speed of passenger cars is measured over a longer stretch of highway, averaged over both directions. Truck speeds are not very sensitive to increases in traffic volume, but traffic volume is the main factor affecting the ATS of passenger cars (*22*). In their investigation of two-lane highway performance measures, Al-Kaisy and Karjala (*6*) argued that ATS of passenger cars may more accurately describe speed reduction due to traffic, as passenger car speeds are more affected by high traffic volumes than heavy vehicle speeds.

The v/c ratio, or degree of capacity utilization, has been used as the main performance measure on two-lane highways in Japan, China, and Denmark (*24–26*). It is important to note that the two-lane expressways in Japan are different from the conventional two-lane highways in the United States and most other countries, in that they have limited access (no at-grade intersections) and a median barrier present in all sections (*24*). Furthermore, the v/c ratio has been used as an additional performance measure in Sweden (*27*). This performance measure was also used for estimating the LOS along with operating speed in the second edition of the HCM (*16*).

Follower density (FD) is another performance measure, which was originally adopted by the South African National Roads Agency more than a decade ago (*12*, *28*) and was later investigated in other studies (*6*, *8*, *9*, *11*, *29*). FD is defined as the product of percent followers (PF) and traffic density; therefore, this measure is derived using two important flow characteristics: traffic flow and density. Again, PF is estimated using time headway, which is a microscopic flow characteristic. FD provides a better indication for warranting highway upgrades and improvements. In addition to HCM performance measures, the Oregon Department of Transportation has used FD in operational analyses of two-lane highways in recent years for the obvious advantage of capturing the effects of platooning and traffic level (*30*). Table 2 shows the LOS criteria as a function of FD for Class I, and Class II highways that were developed by Oregon DOT.

Alternative Performance Measures Proposed in Literature

Given the reported limitations in the current HCM performance measures (*3–14*), several alternative performance measures have been proposed by studies here in the United States and in other countries.

A study by Al-Kaisy and Karjala (*6*) investigated a new speed-related performance measure among other performance measures on two-lane highways. The new measure is calculated as the average travel speed of

Table 2. LOS Criteria for Two-Lane Highways (*30*)

LOS	Class I highways: Follower density (vehicles/mile/lane)	Class II highways: Follower density (vehicles/mile/lane)
A	≤2	≤2.5
B	>2–3.5	>2.5–4.0
C	>3.5–6	>4.0–6.5
D	>6.0–9.0	>6.5–10.0
E	>9.0	>10.0

passenger cars (ATS_{PC}) divided by the free-flow speed of passenger cars (FFS_{PC}) and expressed as a percentage. The researchers argued that the ATS of passenger cars may more accurately describe speed reduction due to traffic, as passenger car speeds are more affected by high traffic volumes than heavy vehicle speeds. Further, using ATS as a percentage of FFS was viewed as a good indicator of the amount of speed reduction due to traffic and the amount of vehicular interaction in the traffic stream. However, evaluations using field data showed that the speed measures did not exhibit good correlations with platooning variables as compared to other performance measures investigated in this study. Later studies by Hashim and Abdel-Wahed (*8*) and Moreno et al. (*9*) reported similar results using field data in Egypt and Spain. In his study on PTSF in Finland, Luttinen reported on another old study by O.K. Normann, who suggested the use of speed differences between successive vehicles on two-lane highways, among other proposed performance measures (*5*).

A couple of studies in the literature proposed the use of passing maneuvers in measuring performance on two-lane highways. The platooning phenomenon on two-lane highways and the associated delay are directly related to passing opportunities and the ability of platoon vehicles to pass slower vehicles and improve their speeds. As such, a few performance measures were proposed for assessing performance on two-lane highways that are related to passing maneuvers.

A study by Morrall and Werner (*31*) proposed the use of an overtaking ratio as a supplementary indicator of the LOS on two-lane highways. This measure is obtained by dividing the number of passes achieved by the number of passes desired. According to the study, the number of passes achieved is the total number of observed passes for a given two-lane highway, whereas the number of passes desired is the total number of passes for a two-lane highway with continuous passing lanes with similar vertical and horizontal geometry. The overtaking ratio considers the effect of the percentage of no-passing zones and passing lanes. The researchers showed that the effect of passing lanes as measured by the overtaking ratio is greater than PTD, used by the 1985 HCM. The overtaking ratio, along with the average number of passes per vehicle, was also proposed by O. K. Normann, as reported by McLean (*32*) and Luttinen (*5*).

Another measure that has been used extensively both in practical applications as well as in published research is time headway, a major microscopic traffic flow characteristic. For various practical reasons, time headway has been used solely for identifying platoons using empirical traffic data and field measurements. One important reason for using time headway is that this measure can readily be extracted from the output of conventional traffic recorders, which have the ability to provide raw data (i.e., timestamp records for individual vehicle arrivals). The second equally important reason is the fact that time headway is a good indicator of the interaction between successive vehicles in the traffic stream and, consequently, in determining the status of a vehicle being in a following mode (i.e., being part of a vehicular platoon).

Most of the proposed headway-related measures involve the estimation of the percentage of headways less than a pre-specified cut-off value, a measure called in recent studies percent followers (PF). The percentage of headways less than 3 s has been used by the HCM as a surrogate measure for PTSF using field data (*1*). A few recent studies have examined this along with other proposed performance measures to evaluate their suitability for use on two-lane highways (*6*, *8*, *9*, *11*, *28–30*). This performance measure does not necessarily reflect the effect of traffic level, that is, there could be a two-lane highway with low traffic volume but still a high percentage of short headways. Therefore, when used alone, this measure could be misleading in highway improvement and upgrade decisions. Various headway cut-off values were proposed to estimate the percentage of vehicles in following mode in the traffic stream (*12*, *13*, *29*).

Besides PF, that is, the percent of short headways in a traffic stream, a few other measures have been proposed in the literature that are associated with more than one traffic stream parameter, and as such, they are referred to in this discussion as compound measures. The merit of using compound measures is the fact that those measures usually combine the advantages of more than one indicator of traffic performance on two-lane highways (e.g., amount of platooning and traffic level). One such measure that was proposed in the literature is follower flow (FF), which was investigated by the South African National Roads Agency (*12*). It is defined as the hourly rate of vehicles in following mode that pass a point along a two-lane highway. This measure can easily be estimated as the product of PF and flow rate. FF was investigated among several other performance measures in the development of the current South African two-lane highway methodology.

Penmetsa et al. (*29*) proposed another compound measure for Indian two-lane highways under mixed traffic conditions. The new measure is called number of followers (NF), which represents the number of vehicles moving in the same lane of two-lane highways with gaps less than 2.6 s. This measure showed a strong correlation with traffic volume. Penmetsa et al. (*29*) argued that NF alone is not able to identify the congestion condition on a highway network. Therefore, number of followers as a proportion of capacity (NFPC) was introduced as a new performance measure. The new measure exhibited a very strong correlation with traffic volume and was found to explain the quality of traffic flow in a better way.

Another compound measure that was proposed in the literature is percent impeded (PI). This measure was originally proposed by Al-Kaisy and Freedman (*33*), and was later investigated by Hashim and Abdel-Wahed (*8*), Ghosh et al. (*34*), and Moreno et al. (*9*). PI is defined as the percentage of vehicles impeded by slower-moving vehicles in a directional traffic stream measured at a point location. PI is calculated as the probability of desired speeds being greater than the average speed of platoon leaders multiplied by the percentage of vehicles with headways less than a pre-specified threshold value. For establishing the distribution of desired speeds, vehicles with headways more than 8 s were used by Al-Kaisy and Freedman (*33*). Platoon leaders were used as a representative sample of slow-moving vehicles. This measure is derived using flow and speed characteristics.

PI was further used to develop other compound measures. Impeded flow which shows the hourly rate of vehicles impeded by slow-moving vehicles, that is, platoon leaders, was suggested by Al-Kaisy et al. (*35*, *36*). It is calculated as the product of PI and the hourly rate of traffic flow. They also proposed impeded density (ID), which refers to the number of vehicles impeded per mile per lane in one direction of travel. It is calculated as the product of PI and traffic density. These measures along with other performance measures were investigated by Al-Kaisy et al. (*35*, *36*) using field data from two-lane highways in Montana, Idaho, Oregon, and North Carolina. Among all compound measures, FF followed by FD exhibited highest correlations with traffic variables.

Freedom of flow is another performance measure proposed for use on two-lane highways (*37*). Polus and Cohen (*37*) used the M/M/1 queuing to model vehicular platoons on two-lane highways. In their approach, a platoon is considered as a queue system with random arrivals (vehicles joining the back of queue), random departures (vehicles passing platoon leader), and a single processing channel. To identify platooned vehicles, the researchers used a 3-s follower headway threshold. In their research, two proposed traffic parameters were introduced: traffic intensity and freedom of flow. Similar to queuing analysis, traffic intensity (ρ) in the context of platoon analysis refers to the ratio between the average time spent in the first position when waiting for an appropriate gap and the average inter-arrival times at the back of the queue. On the other hand, freedom of flow (η) refers to the ratio between the time of undisturbed driving (between platoons) and the time interval in which the driver is in the first position behind a slower-moving vehicle while waiting to pass. This ratio may reflect the perception of tension experienced by drivers in a more faithful way. Freedom of flow can be estimated using Equation 1 below.

$$\eta = \frac{N_o}{\rho} \tag{1}$$

where

η = Freedom of flow,
ρ = Traffic intensity, and
N_o = Average number of headways between platoons.

Preliminary Assessment of Performance Measures

In this section, an initial qualitative assessment of the surveyed alternative performance measures is presented. The HCM performance measures will also be included in this assessment to help in understanding the merits of the proposed alternative measures. As stated before, it is desirable for prospective performance measures on two-lane highways to meet the set of criteria discussed earlier in this study. Table 3 summarizes all performance measures discussed in this study and the degree to which each measure satisfies the aforementioned six criteria. In this subjective assessment, each performance measure is evaluated against the six criteria independently (i.e., assessed assuming it is used solely for measuring performance on two-lane highways). It is important to remember that those criteria do not have the same significance when it comes to conducting two-lane highway operational analyses. For example, to easily measure performance in the field may carry more weight than supporting environmental impact analysis. As the objective of this research is to examine performance measures rather than to recommend a particular measure for operational analysis on two-lane highways, no weights were assigned to the six evaluation criteria.

At a glance, it can be clearly shown that no single measure largely satisfied all evaluation criteria and that each measure satisfied some criteria to a higher degree but scored low on other criteria. Though this is the case for almost all performance measures included in this evaluation, the following observations can be made:

1. The two major HCM performance measures on two-lane highways, PTSF and ATS, do not reasonably satisfy most of the six evaluation criteria.

Table 3. Preliminary Assessment Matrix of Performance Measures on Two-Lane Highways

Performance measure	(1) Road user perception	(2) Easy to measure	(3) Sensitive to road conditions	(4) Compatible with other facilities	(5) Describes all flow regimes	(6) Support other analyses			
						Safety	Economic	Environmental	Reliability
HCM–PTSF[a]	XXX[b]	X	X	X	X[c]	X	X	X	X
HCM–ATS	X	XXX	X	XX	X[c]	X	XX	X	X
HCM–(ATS/FFS)	XX	XXX	XX	X	XX[c]	X	XXX	XX	X
ATS_{PC}	X	XXX	X	XX	X	X	XX	XX	X
ATS_{PC} as a percentage of FFS of PCs (ATS_{PC}/FFS_{PC})	XX	XXX	XX	X	XX	X	XXX	XX	X
Speed variance	X	XXX	X	X	X	XX	X	X	X
Demand-to-capacity (d/c) ratio	XX	XXX	X	X	XXX	XX	XX	XX	XX
Traffic density	X	XXX	XX	XXX	XXX	XX	X	X	XX
Overtaking ratio	X	X	XX	X	X	X	X	X	X
Average # of passes per vehicle	X	X	XX	X	X	XX	X	X	X
Percent followers (PF)	XX	XXX	XX	X	X	X	X	X	X
Follower flow (FF)	X	XXX	XXX	X	X	XX	X	X	X
Follower density (FD)	XX	XXX	XXX	XX	X	XX	X	X	X
Percent impeded (PI)	XXX	XX	XXX	X	X	XX	X	X	X
Impeded flow (IF)	X	XX	XXX	X	X	XX	X	X	X
Impeded density (ID)	XX	XX	XXX	XX	X	XX	X	X	X
Freedom of flow	X	X	XX	X	XX	XX	X	X	X

[a]PTSF is estimated using flow rates in the analytical methodology and using percent followers (PF) as a surrogate measure for field estimation.
[b]X = hardly meeting criterion; XX = moderately meeting criterion; XXX = largely meeting criterion.
[c]Level of service F is not defined for any of the three HCM 2016 performance measures.

2. Aside from being easy to measure in the field, speed-related measures scored generally low on all other criteria except for supporting economic and (to a lesser degree) environmental analyses.
3. Performance measures based on passing maneuvers scored notably low on most of the evaluation criteria.
4. Compound measures involving headway and either density or flow rate showed a higher degree of satisfying the six criteria particularly FD and ID.

The above discussion suggests that coming up with a single performance measure that satisfactorily meets all criteria may prove to be very difficult, and therefore, it may be advantageous to use more than one performance measure in describing performance on two-lane highways.

Summary

The current study presented an overview of performance measures that have been used in practice or proposed in the literature for assessing performance on rural two-lane highways. A historical background on two-lane highway performance measures used by the HCM since its first edition was presented. Furthermore, the most important criteria for effective performance measures were presented and discussed. The review involved the performance measures currently used by the HCM, those that have been used in other countries, and other alternative performance measures proposed in the literature. The study concluded by assessing the degree each performance measure satisfies each of the criteria identified for effective performance measures. Though subjective in nature, this preliminary assessment provides valuable information about the merits, or lack thereof, of a large number of performance measures identified in this study. Results suggest that no single measure largely satisfied all evaluation criteria and that each measure satisfied some criteria to a higher degree but scored low on other criteria. Therefore, coming up with a single performance measure that satisfactorily meets all criteria may prove to be very difficult, and it may be advantageous to use more than one performance measure in describing performance on two-lane highways.

Acknowledgments

The authors would like to acknowledge the financial support provided for this study by the National Cooperative Highway Research Program.

Author Contributions

The authors confirm contribution to the paper as follows: study conception and design: A.A-K. and S.W.; data collection: A.J., S.W., and T.L.; draft manuscript preparation: A.A-K., S.W., T.L., and R.D. All authors reviewed the results and approved the final version of the manuscript.

References

1. *Highway Capacity Manual: A Guide for Multimodal Mobility Analysis*, 6th ed. Transportation Research Board, Washington, D.C., 2016.
2. Washburn, S., S. McLeod, and G. K. Courage. Adaptation of Highway Capacity Manual 2000 for Planning-Level Analysis of Two-Lane and Multi-Lane Highways in Florida. *Transportation Research Record: Journal of the Transportation Research Board*, 2002. 1802: 62–68.
3. Al-Kaisy, A., and Z. Freedman. Empirical Examination of Passing Lane Operational Benefits on Rural Two-Lane Highways. *Journal of Transportation Research Forum*, Vol. 49, No. 3, 2010, pp. 53–68.
4. Dixon, M. P., S. S. K. Sarepali, and K. A. Young. Field Evaluation of Highway Capacity Manual 2000 Analysis Procedures for Two-Lane Highways. *Transportation Research Record: Journal of the Transportation Research Board*, 2002. 1802: 125–132.
5. Luttinen, T. Percent Time-Spent-Following as a Performance Measure for Two-Lane Highways. *Transportation Research Record, Journal of the Transportation Research Board*, 2001. 1776: 52–59.
6. Al-Kaisy, A., and S. Karjala. Indicators of Performance on Two-Lane Rural Highways: An Empirical Investigation. *Transportation Research Record: Journal of the Transportation Research Board*, 2008. 2071: 87–97.
7. Al-Kaisy, A. F., and C. Durbin. Estimating Percent Time Spent Following on Two-Lane Highways: Field Evaluation of New Methodologies. Presented at 86th Annual Meeting of the Transportation Research Board, Washington, D.C., 2007.
8. Hashim, I., and T. A. Abdel-Wahed. Evaluation of Performance Measures for Rural Two-Lane Roads in Egypt. *Alexandria Engineering Journal*, Vol. 50, No. 3 2011, pp. 245–255.
9. Moreno, A., C. Llorca, T. Sayed, and A. Garcia. Field Evaluation of Traffic Performance Measures for Two-Lane Highways in Spain. Presented at 93rd Annual Meeting, Transportation Research Board, Washington, D.C., 2014.
10. Romana, M. G., and I. Pérez. Measures of Effectiveness for Level-of-Service Assessment of Two-Lane Roads: An Alternative Proposal Using a Threshold Speed. *Transportation Research Record: Journal of the Transportation Research Board*, 2006. 1988: 56–62.
11. Catbagan, J., and H. Nakamura. Evaluation of Performance Measures for Two-Lane Expressways in Japan. *Transportation Research Record, Journal of the Transportation Research Board*, 2006. 1988: 111–118.
12. Van As, C. *The Development of an Analysis Method for the Determination of Level of Service on Two-Lane Undivided Highways in South Africa: Project Summary*. South African National Roads Agency, Limited, Pretoria, 2003.
13. Al-Kaisy, A., A. Jafari, and S. Washburn. Following Status and Percent Followers on Two-Lane Two-Way Highways:

Empirical Investigation. Presented at 96th Annual Meeting of the Transportation Research Board, Washington, D.C., 2017.

14. Ibrahim, M., and O. C. Puan. Measuring Driver's Percent-Time-Spent-Following Using a Moving Video Recording Technique. *Journal of Applied Mechanics and Materials*, Vol. 567, pp. 736–741.
15. *Highway Capacity Manual*. Highway Research Board and Bureau of Public Roads, Washington, D.C., 1950.
16. *Special Report 87: Highway Capacity Manual*, 2nd ed. Highway Research Board, Washington, D.C., 1965.
17. *Special Report 209: Highway Capacity Manual*, 3rd ed. (1997 update). TRB, National Research Council, Washington, D.C., 1985.
18. *Highway Capacity Manual*. TRB, National Research Council, Washington, D.C., 2000.
19. *Highway Capacity Manual 2010*. Transportation Research Board of the National Academies, Washington, D.C., 2010.
20. Bessa, J. E., and J. R. Setti. *Deviation of ATS and PTSF Functions for Two-lane, Rural Highways in Brazil*. Presented at the 6th International Symposium on Highway Capacity and Quality of Service, Stockholm, Sweden, 2011.
21. Al-Kaisy, A., A. Jafari, and S. Washburn. Performance Measures on Two-Lane Highways: Survey of Practice. Presented at 95th Annual Meeting of the Transportation Research Board, Washington, D.C., 2016.
22. Brilon, W., and F. Weiser. Two-Lane Rural Highways: The German Experience. *Transportation Research Record: Journal of the Transportation Research Board*, 2006. 1988: 38–47.
23. HBS 2001. *Handbuch fuer die Bemessung von Strassen* (German Highway Capacity Manual). German Road and Transportation Research Association, Cologne, 2005.
24. Nakamura, H., and T. Oguchi. An Investigation on Performance Measures for Two-Lane Expressways in Japan. *Proc., 5th International Symposium on Highway Capacity and Quality of Service*, Technical Papers, Vol. 2, Japan Society of Traffic Engineers, Tokyo, 2006, pp. 87–104.
25. Rong, J., J. Guo, and R. Zhou Chinese Highway Capacity Research Studies and Applications. In *6th International Symposium on Highway Capacity and Quality of Service – Country Reports* (Koutsopoulos, H. N., and K. L. Bang eds.), TRB and KTH, Stockholm, 2011, pp. 52–66.
26. Vejdirektoratet. Kapacitet og serviceniveau. Vejdirektoratet, Vejregelrådet. [Handbook for capacity and service level. Road Directorate, Road Council, Copengagen (in Danish). 2015.
27. Trafikverket. TRVMB Kapacitet och framkomlighetseffekter [TRVMB Capacity and Accessibility Effects]. *TRV*, Vol. 2013, 2014, 64343.
28. Van As, C., and A. Niekerk. The Operational Analysis of Two-Lane Rural Highways. *Proc., 23rd Annual Southern African Transport Conference*, Pretoria, South Africa, 2004.
29. Penmetsa, P., I. Ghosh, and S. Chandra. Evaluation of Performance Measures for Two-Lane Intercity Highways under Mixed Traffic Conditions. *Journal of Transportation Engineering*, Vol. 141, No. 10, 2015, pp. 1–7.
30. Oregon Department of Transportation. *Modeling Performance Indicators on Two-Lane Rural Highways: The Oregon Experience*. http://www.oregon.gov/odot/td/tp/docs/reports/modelingperf.pdf. Accessed December 11, 2014.
31. Morrall, J. F., and A. Werner. Measuring Level of Service of Two-Lane Highways by Overtakings. *Transportation Research Record: Journal of the Transportation Research Board*, 1990. 1287: 62–69.
32. McLean, J. *Two-Lane Highway Traffic Operations: Theory and Practice. Transportation Studies*, Vol. 11. New York: Gordon and Breach Science Publishers, 1989.
33. Al-Kaisy, A., and Z. Freedman. Estimating Performance on Two-Lane Highways: Case Study Validation of a New Methodology. *Transportation Research Record: Journal of the Transportation Research Board*, 2011. 2173: 72–79.
34. Ghosh, I., S. Chandra, and A. Boora. Operational Performance Measures for Two-Lane Roads: An Assessment of Methodological Alternatives. *Proc., 2nd Conference of Transportation Research Group of India (2nd CTRG)*, Agra, India, Procedia - Social and Behavioral Sciences, Vol. 104, Elsevier, December 12–15, 2013, pp. 440–448.
35. Al-Kaisy, A., A. Jafari, and S. Washburn. Measuring Performance on Two-Lane Highways: Empirical Investigation. Presented at 96th Annual Meeting of the Transportation Research Board, Washington, D.C., 2017.
36. Al-Kaisy, A., A. Jafari, and S. Washburn. Measuring Performance on Two-Lane Highways: Empirical Investigation. *Transportation Research Record: Journal of the Transportation Research Board*, 2017. 2615: 62–72.
37. Polus, A., and M. Cohen. Theoretical and Empirical Relationships for the Quality of Flow and for a New Level of Service on Two-lane Highways. *Journal of Transportation Engineering*, Vol. 135, No. 6, 2009, pp. 380–385.

The Standing Committee on Highway Capacity and Quality of Service (AHB40) peer-reviewed this paper (18-04914).

Article

TRR
JOURNAL OF THE TRANSPORTATION RESEARCH BOARD

Transportation Research Record
2018, Vol. 2672(15) 75–84

DOI: 10.1177/0361198118777602
journals.sagepub.com/home/trr
SAGE

Comparison of Highway Capacity Estimation Methods

Mohamadamin Asgharzadeh[1] and Alexandra Kondyli[1]

Abstract

The capacity of a freeway segment is a critical factor for planning, design, and operational analysis of freeway facilities. This research aimed to perform a comparison among well-known freeway capacity estimation methods in order to investigate their application, as well as their advantages and disadvantages. Single estimate capacity methods such as the Van Aerde method, and breakdown probability methods, such as the product limit method (PLM), the Highway Capacity Manual (HCM) method, and the sustainable flow index (SFI) method, were applied at six merge bottleneck locations in the Kansas City area. The results from all methods were compared and the advantages and the disadvantages of each method were discussed. The HCM results showed a significant variability in the estimated breakdown probability function and the resulting capacities. The HCM method was also found to be sensitive to the breakdown probability ratios as a single breakdown observation can significantly shift the fitted distribution and the corresponding capacity estimate. The PLM model provided the highest capacity estimates, followed by the Van Aerde model capacities. The Van Aerde capacities were also found to be closer to the average pre-breakdown flow rates. Finally, the PLM and the SFI method showed consistent performance in comparison to the remaining methods, and flexibility in being applied on different sites with various characteristics.

Freeway capacity is defined as the maximum number of vehicles that can flow through a specific freeway segment. A significant amount of research has defined freeway capacity as a function of the occurrence of the "breakdown" phenomenon, which reflects the transition from uncongested to congested flow regime. This transition is typically manifested by a sharp speed drop or occupancy increase (*1*). These breakdown events usually occur at freeway bottleneck locations, such as merging, diverging, weaving, or lane drop segments, because of increased lane changing activity and turbulence caused by interactions between the mainline and the ramp vehicles. The Highway Capacity Manual 6th edition (HCM6) defines freeway capacity as "maximum flow associated with the occurrence of some type of breakdown, which results in lower speeds and higher densities following the breakdown event" (*2*). The HCM6 also acknowledges that capacity is random in nature, and it may vary from day to day, and as such, the manual includes a probabilistic method for calculating capacity at bottleneck locations. However, specific capacity values at merge segments are not provided in the HCM.

Several researchers have investigated the breakdown occurrence and associated capacity at merge bottlenecks. At these sites, capacity is usually defined as the pre-breakdown flow rate, that is, the flow rate right before the breakdown occurrence (*2*). Pre-breakdown flows have significant variation at the same site and from day to day. The main contributors to the capacity variation are driver behavior, occurrence of incidents, work zones, adverse weather, and spillbacks from downstream bottlenecks.

Earlier attempts to obtain freeway capacities considered capacity as a single constant value. The first publication on this subject belongs to Greenshields et al. (*3*) who suggested that the maximum capacity of the freeway is equal to the volume that corresponds to 50% of the free-flow speed (FFS). Greenshields et al. (*3*) developed a symmetrical speed-flow diagram assuming that the speed–density relationship follows a linear trend. Van Aerde proposed a mathematical model to estimate capacity as a single-point value (*4*). Van Aerde's model was a more complicated form of Greenshields et al.'s model by taking vehicle headway parameters into account. The estimated capacity in this model was the point with the highest flow rate in the speed-flow diagram. However,

[1]Department of Civil, Environmental and Architectural Engineering, University of Kansas, Lawrence, KS

Corresponding Author:
Address correspondence to Alexandra Kondyli: akondyli@ku.edu

freeway capacity varies from day to day under similar prevailing conditions; therefore, a single value cannot fully represent capacity (*1*, *5*, *6*). This research implemented the Van Aerde model as the only single capacity estimate model because of its simplicity and convenient application, and compared it with probabilistic-based capacity estimation models.

Contrary to single-point capacity estimate models, breakdown probability models provide a distribution function based on observed breakdowns to account for capacity variability (*7*). Hall and Agyemang-Duah (*7*) and Minderhoud et al. (*8*) discussed the capacity drop before and after breakdown and applied the normal distribution on the pre-breakdown flow rates. Other researchers have also used the PLM, which is based on lifetime data analysis of breakdowns and considers both breakdown flows and uncongested observations (*9*, *10*). The PLM is a non-parametric method that provides the probability of a breakdown at a certain volume (i.e. capacity). Past research has found that the Weibull distribution provides an acceptable fit to the PLM (*10*). A significant amount of data (6 months to 1 year) is typically required to build valid probability models using the PLM.

Many researchers have also investigated the selection of a single acceptable capacity value obtained from the distribution of the pre-breakdown flow rates. Brilon (*11*) introduced a term called "efficiency," which is the product of volume, average speed, and duration of the analysis period, and he defined capacity as the maximum achievable flow in a section. Brilon (*11*) suggested that the highest efficiency of a freeway is approximately at 90% of the 1-H average capacity as obtained from the PLM. A recent probabilistic approach estimated capacity by maximizing the sustainable flow index (SFI) parameter, which is the product of the volume and the corresponding survival probability (*12*). The SFI method works similarly to the PLM, particularly in fitting the Weibull distribution on the uncongested portion of speed-flow data.

Recently, the HCM6 introduced a probabilistic capacity estimation method using the breakdown probability ratio (number of breakdowns divided by total number of observations in each volume interval [*2*]). Similar to the PLM, the HCM6 method considers both congested and uncongested observations to develop the proposed distribution, and requires intensive data preparation including data aggregation and breakdown identification. Since the HCM6 method is relatively new, its application to estimate freeway capacity has been limited.

To estimate capacity based on the probability of breakdown, one must first define and identify the breakdown occurrences. Most of the previous studies have used macroscopic speed drops downstream of the merge bottleneck as the breakdown criterion (*1*, *6*, *10*, *13–15*). According to the HCM6 method, breakdown occurs when speed falls below 75% of the FFS for at least 15 min (*2*). Brilon et al. (*2*) defined breakdown based on a speed drop greater than 10 km/h (6 mph) and below the threshold of 72 km/h (45 mph) over 5 min. Kondyli et al. (*1*) used a 10 mph (16 km/h) speed drop sustained over a 10-min period.

The main objective of this paper is to examine and compare the four freeway capacity estimation methods discussed earlier (Van Aerde, PLM, SFI, and HCM6) using data obtained at several active merge bottlenecks in the Kansas City metropolitan area. To accomplish this objective, historic macroscopic traffic data were obtained at several merge bottleneck locations in the Kansas City area for a 1-year period. To apply the selected capacity estimation algorithms, only the breakdowns that were caused because of merging operations and not because of downstream queue spillback, incidents, or adverse weather events were considered. The final data set was used to estimate capacities using the four methodologies. A comparison of the capacities estimated by these methodologies, along with their advantages and disadvantages, is also presented. The findings of this research could assist transportation professionals and researchers in selecting the appropriate capacity estimation method.

Methodology

The macroscopic traffic data used in this research include average speeds, flows, and occupancies per lane aggregated in 5-min intervals. The traffic data were obtained from the KC Scout database (www.kcscout.com) for a 12-month period (January–December 2016). The study focused on breakdown events due to merging operations; therefore, incident or adverse weather-induced breakdown events were excluded from the analysis. Incident data along major corridors in the Kansas City area were also obtained from the KC Scout database. If an incident occurred within 1 mi downstream of the merge junction, and before the recorded speed drop at the merge detector, then it was assumed that the incident contributed to the breakdown, and that breakdown event was excluded from the analysis. Similarly, any records of snow or precipitation greater than 0.20 in. within 1 hour before the recorded congestion were removed from further analysis. Historical weather-related data were obtained through the https://www.wunderground.com website.

The HCM6 method was used to identify the breakdown events to use in the three breakdown-based capacity estimation methodologies. According to this method, breakdowns occur when the average speed falls below 75% of the FFS for at least 15 min (*2*). It is also assumed that traffic conditions recover when the average speed increases to the recovery speed (90% of FFS). It is further assumed that FFS is equal to the average speed during free-flowing conditions [e.g. when volume is less than 600 vehicles per hour per lane (vphpl)].

To distinguish between true breakdowns and queue spillbacks from downstream congestion, sensor data further downstream of the merge were analyzed. If congestion was first detected at the merge and (near) free-flowing conditions occurred at the downstream location, then it was assumed that the breakdown event was a true breakdown. If congestion was first identified downstream of the merge, then it was assumed that congestion propagated from downstream, and therefore, that event was removed from further analysis.

Van Aerde Method

The Van Aerde method requires fitting a curve to estimate the relationship in the speed-flow diagram. Equation 1 shows the speed-volume curve suggested by Van Aerde (*4*).

$$q_i = \frac{u_i}{c_1 + \frac{c_2}{\mathrm{FFS} - u_i} + c_3 u_i} \tag{1}$$

Where:

u_i = speed at interval i for all observed data in both congested and uncongested regimes;

q_i = volume at interval i for all observed data in both congested and uncongested regimes;

FFS = free-flow speed; and

c_1, c_2, and c_3 = first, fixed, and second headway constants coefficient, respectively.

In our research, a metaheuristic algorithm based on Particle Swam Optimization (*16*), was implemented to minimize Z, which is the sum of normalized squared differences between the observed and estimated speed, density, and volume to obtain the four unknown parameters (jam density, speed at capacity, maximum volume, and FFS). The objective function is shown in Equation 2.

$$\text{Min Z} = \sum_i \left[\left(\frac{u_i - \hat{u}_i}{\bar{u}} \right)^2 + \left(\frac{q_i - \hat{q}_i}{\bar{q}} \right)^2 + \left(\frac{d_i - \hat{d}_i}{\bar{d}} \right)^2 \right] \tag{2}$$

Where:

$\hat{u}_i$, $\hat{q}_i$, and $\hat{d}_i$ = estimated speed, volume, and density, respectively; and

$\bar{u}$, $\bar{q}$, and $\bar{d}$ = average speed, volume, and density, respectively.

Product Limit Method

The PLM is based on lifetime data analysis research that is used for estimating the time until failure of mechanical parts or the duration of human life (*9*). Here, the event of a failure at time t is the event of a breakdown at volume q, and the lifetime is the maximum pre-breakdown volume. The breakdown volume (q) is assumed to be the volume that triggers the abrupt speed drop. The non-parametric PLM (*17*) is used to obtain the distribution function of the breakdown volume according to the following equation (*10*).

$$F_c(q) = 1 - \prod_{i:q_j<q} \frac{k_i - 1}{k_i} \quad i \in \{B\} \tag{3}$$

Where:

$F_c(q)$ = cumulative capacity distribution function;

q = observed traffic volume vphpl);

$q(i)$ = traffic volume in time interval i (pre-breakdown volume);

k_i = number of intervals with $q \geq q_i$; and

$\{B\}$ = set of breakdown intervals (5-min observations).

Once the breakdown probability models at each site were constructed using the PLM, the Weibull distribution was fitted and the distribution parameters were estimated.

Sustainable Flow Index

Recently, Shojaat et al. (*12*) proposed the SFI method in extension to the PLM. This method utilizes the breakdown probability function described in Equation 3, and calculates the SFI according to Equation 4.

$$\mathrm{SFI} = q_i \cdot (1 - F_c(q_i)) \tag{4}$$

Equation 4 indicates that the SFI at each volume (q_i) is the product of the volume and its survival function. This method defines capacity as the volume at which SFI is maximized. If the breakdown probability function is assumed to be Weibull distributed, then the optimum flow rate (capacity) for which the SFI is maximized is:

$$q_{opt} = \beta \cdot \left(\frac{1}{\alpha} \right)^{1/\alpha} \tag{5}$$

Where β and α are the shape and scale parameters of the fitted Weibull distribution.

HCM6 Capacity Estimation Method

The HCM6 method also requires breakdown probability analysis at each site. The same procedure was followed to identify breakdowns and recoveries throughout the entire data set. To apply this method, 5-min data were aggregated to 15-min data, as per HCM's requirements. The pre-breakdown and breakdown volume observations were grouped into volume intervals (bins) of 100 vphpl. After the total number of pre-breakdown

Table 1. Description of Data Collection Sites

ID	Location	Number of lanes	Speed limit (mph)	FFS (mph)	Breakdown speed (mph)	Total number of breakdowns	Number of true breakdowns	Peak period
1	I-35 SB South of 67th Street	4	60	62.9	47.2	458	258	6–9 a.m. 2–7 p.m.
2	I-35 SB South of 75th Street	4	60	66.1	49.6	461	308	6–9 a.m. 2–7 p.m.
3	I-35 NB North of 75th Street	4	60	60.6	45.5	107	23	6–9 a.m. 3–6 p.m.
4	I-70 EB East of Blue Ridge	3	65	68.3	51.2	239	104	2–7 p.m.
5	I-70 WB East of US-40	3	65	68.8	51.6	248	22	6–9 a.m.
6	I-70 WB East of Noland Road	3	65	65.1	48.8	258	134	6–8 a.m.
7	I-435 EB East of 104th Street	4	65	70.0	52.5	90	56	6–9 a.m. 2–6 p.m.
8	I-435 EB East of Holmes Road	4	65	68.9	51.7	280	119	3–6 p.m.

observations was acquired within each bin, the breakdown probability was calculated by dividing the number of pre-breakdown flow rates by the total number of observations in each bin. The calculated breakdown probabilities were then used to fit the Weibull distribution. The estimated parameters (shape and scale) from the Weibull distribution were used in Equation 6 to estimate a single capacity value based on the acceptable breakdown probability rate, λ.

$$\text{Capacity} = \beta \cdot \sqrt[\alpha]{-\ln(1-\lambda)} \tag{6}$$

Where:

λ is the acceptable probability rate (HCM6 suggested value: $\lambda = 0.15$);

α and β as defined previously.

Results and Discussion

The study sites analyzed in this paper included multiple freeway merging junctions in the Kansas City metro area. All sites were experiencing recurring congestion because of merging operations. Following the HCM6 guidance (*2*), the breakdown events were identified when the average speed dropped below a certain speed threshold, which was a function of the FFS of each segment.

Table 1 summarizes the eight sites that were considered in the analysis, along with pertinent information on geometric and operational conditions obtained from analyzing data for 1 year. Table 1 also includes the breakdown speeds that were estimated for each site.

In Table 1, the column with the total number of breakdowns (column 7) includes all observations of congestion due to merging operations, incidents, adverse weather, or queue spillbacks from downstream. The column with the true number of breakdowns (column 8) includes only observations with demand-induced congestion due to merging operations, which were eventually used for developing the capacity estimation models. Although all eight sites were active bottlenecks (i.e. they experience true breakdowns), the number of true breakdowns at Sites 3 and 5 was limited (23 and 22, respectively), and it would not be possible to construct a breakdown probability model with so few data points. As such, these two sited were removed from further analysis.

The HCM6, Van Aerde, and PLM models were applied to the selected data set. Figure 1 shows the field measured speed-flow data and the Van Aerde curve. In the Van Aerde model, the highest volume on the fitted speed-flow curve is assumed to be the capacity. Figure 1 shows the speed-flow graphs and the fitted Van Aerde curves for all merge junctions analyzed. This figure also captures differences among the various sites. For example, Sites 2 and 7 do not fully capture the congested regime, possibly because the detector is located further downstream of the merge; therefore, queuing conditions are not observed there.

The HCM6 breakdown probability and fitted Weibull curve, and the PLM Weibull-fitted curve for all merge junctions analyzed are presented in Figure 2. For the HCM6 model, the estimated parameters from the fitted Weibull distribution were used in Equation 6 to calculate the proposed capacity value.

According to Figure 2, the pre-breakdown flow rate observations start at somewhere around 1,300 vphpl and continue to 2,200 vphpl. Also, at low breakdown probabilities, the PLM Weibull-fitted curve is shifted to the right compared to the HCM6 Weibull-fitted curve. As such, the PLM predicts slightly higher capacities compared to the HCM6 method, assuming the selected breakdown threshold is 15% ($\lambda = 0.15$) for both models. In addition, the ratio of the total number of breakdowns

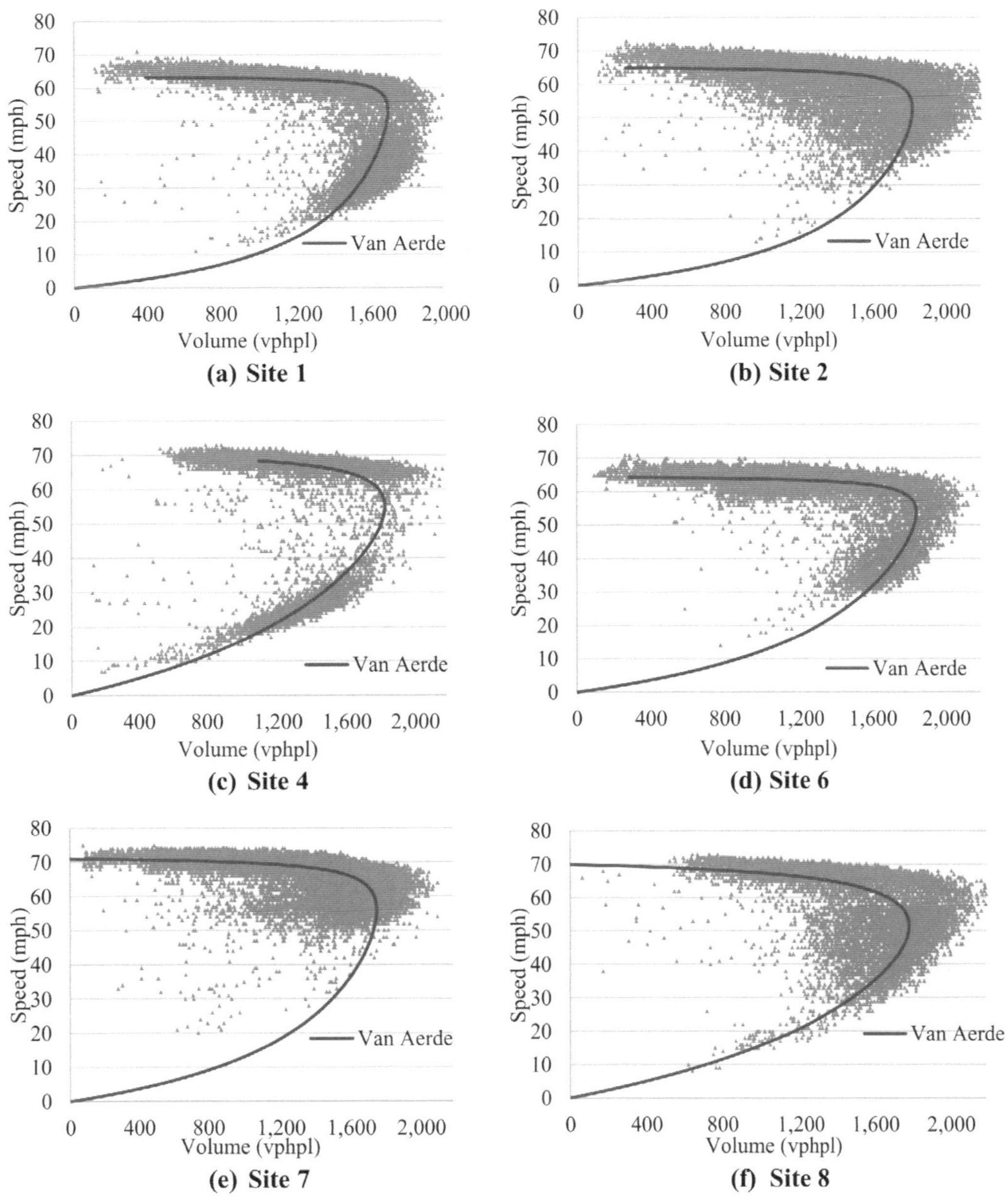

Figure 1. Speed-flow observations and Van Aerde application at (*a*) I-35 SB South of 67th Street, (*b*) I-35 SB South of 75th Street, (*c*) I-70 EB East of Blue Ridge, (*d*) I-70 WB East of Noland Road, (*e*) I-435 EB East of 104th Street, and (*f*) I-435 EB East of Holmes Road.

to the number of observations at each bin needs to be high enough so that the HCM6 Weibull distribution could fit. Similarly, the PLM requires a sufficient number of breakdown observations to produce an acceptable model. Both PLM and HCM6 Weibull distributions were fitted for all sites except for Site 7 (Table 2) because of the limited number of breakdown observations and low probability ratios at that site.

To compare the PLM Weibull-fitted distribution with the remaining capacity estimation methods, the 15th and 20th percentiles of the cumulative fitted distribution were chosen as the capacity values. The derived PLM Weibull distribution parameters were also used to calculate the highest SFI and obtain the capacity values based on this method. Table 2 shows the capacity results based on the average pre-breakdown flow rate, the 15th and 20th

Table 2. Capacity Results Summary

ID	Location	Average pre-breakdown flow (vphpl)	PLM Weibull 15% (vphpl)	PLM Weibull 20% (vphpl)	SFI (vphpl)	Van Aerde (vphpl)	HCM6 (vphpl)
1	I-35 SB South of 67th Street	1,679	1,754	1,779	1,655	1,707	1,538
2	I-35 SB South of 75th Street	1,832	1,991	2,042	1,881	1,830	1,722
4	I-70 EB East of Blue Ridge	1,851	1,935	1,968	1,823	1,840	1,812
6	I-70 EB East of Noland Road	1,857	1,919	1,948	1,810	1,855	1,800
7	I-435 EB East of 104th Street	1,571	NA	NA	NA	1,761	NA
8	I-435 EB East of Holmes Road	1,803	1,928	1,970	1,817	1,792	1,925

Note: NA = not applicable.

Table 3. HCM Evaluation at Site 1 (I-35 SB South of 67th Street)

		Original data			Modified data		
Bin #	Volume range (vphpl)	Number of observations	Number of breakdowns	Breakdown probability	Number of observations	Number of breakdowns	Breakdown probability
1	100–200	36	0	0.000	36	0	0.000
2	200–300	450	0	0.000	450	0	0.000
3	300–400	629	0	0.000	629	0	0.000
4	400–500	625	0	0.000	625	0	0.000
5	500–600	770	0	0.000	770	0	0.000
6	600–700	860	0	0.000	860	0	0.000
7	700–800	809	0	0.000	809	0	0.000
8	800–900	511	0	0.000	511	0	0.000
9	900–1,000	622	0	0.000	622	0	0.000
10	1,000–1,100	1,402	4	0.003	1,402	4	0.003
11	1,100–1,200	1,857	6	0.003	1,857	6	0.003
12	1,200–1,300	1,192	5	0.004	1,192	5	0.004
13	1,300–1,400	782	8	0.010	782	8	0.010
14	1,400–1,500	684	25	0.037	684	25	0.037
15	1,500–1,600	458	74	0.162	458	74	0.162
16	1,600–1,700	329	98	0.298	329	98	0.298
17	1,700–1,800	130	53	0.408	130	53	0.408
18	1,800–1,900	30	10	0.333	30	10	0.333
19	1,900–2,000	4	0	0.000	5	1	0.200
20	2,000–2,100	0	0	0.000	0	0	0.000

percentiles of the PLM distribution, the SFI method, the Van Aerde method, and the HCM6 method.

Based on the results in Table 2, the Van Aerde produced higher capacity estimates compared to the SFI and the HCM6, as this method is based on maximum flows that do not necessarily coincide with the observed breakdown flows. The Van Aerde method also gave the closest estimates to the average pre-breakdown flow rates compared with the other methods. With the exception of the Van Aerde method, all remaining methods define capacity as a function of the breakdown events, and as such, the resulting capacities are significantly lower (excluding the PLM capacities).

In addition, the average pre-breakdown flow takes into account only the flow rates before the breakdown events, but the remaining probabilistic models (PLM, SFI, HCM6) consider both pre-breakdown and uncongested flow rates. Table 2 also shows that the 15th and 20th percentile PLM Weibull capacities were greater than the average pre-breakdown flows by about 5.6% and 7.6%, respectively. On average, the SFI method also resulted in lower capacities compared to the PLM, the Van Aerde, or the average pre-breakdown flow rates; however, this can be explained by the fact that the optimum volume according to the SFI method is defined as "the theoretical average volume that is sustained *without* a traffic breakdown" (*12*). As such, the capacity obtained from the SFI method corresponds to less than the 10th percentile of the PLM Weibull-fitted distribution.

The capacity volumes resulted from the HCM6 method were by far the lowest capacity estimates among all

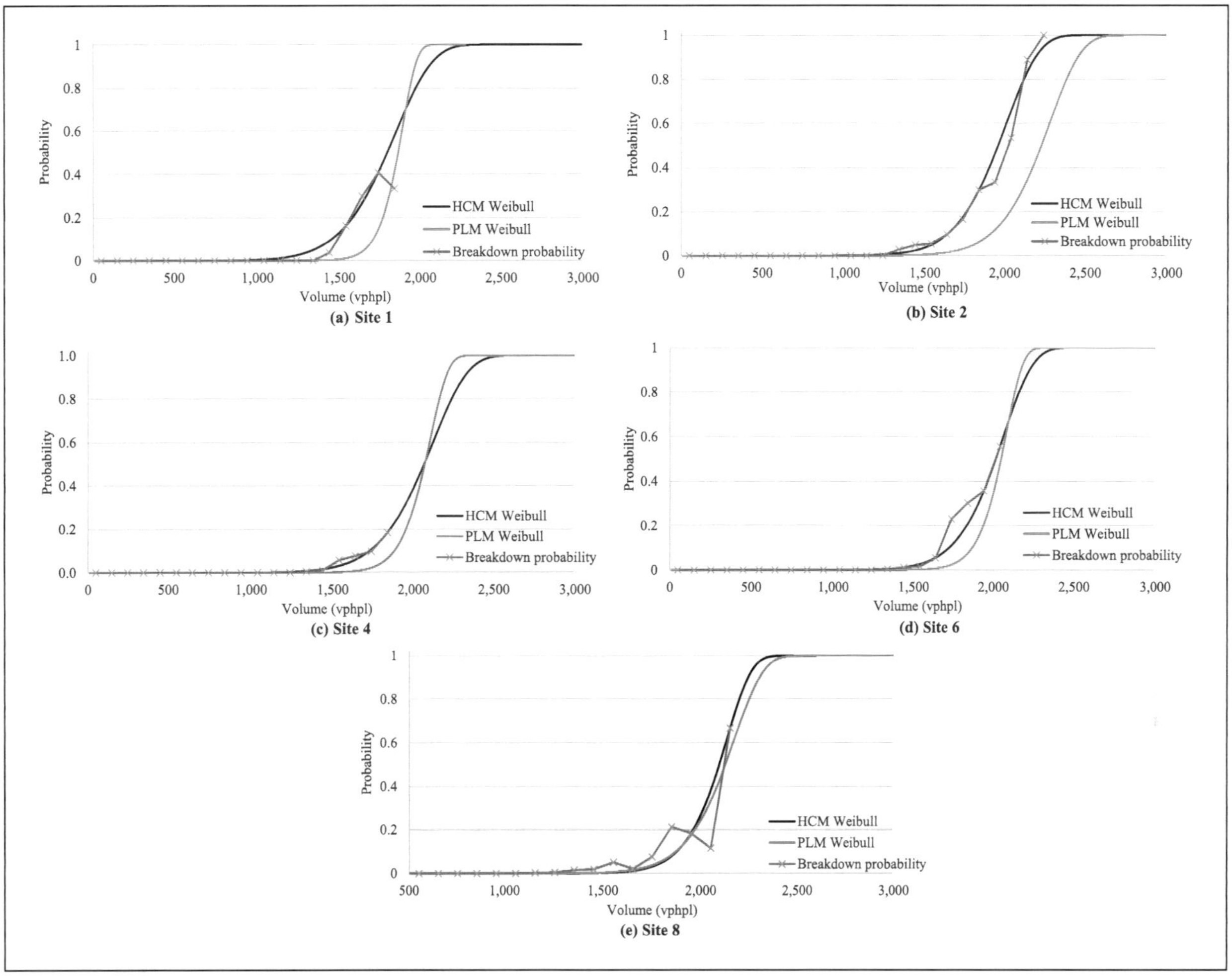

Figure 2. HCM6 and PLM application at (*a*) I-35 SB South of 67th Street, (*b*) I-35 SB South of 75th Street, (*c*) I-70 EB East of Blue Ridge, (*d*) I-70 WB East of Noland Road, and (e) I-435 EB East of Holmes Road.

methods and percentiles used in most cases, and they displayed significant variability across the different sites. According to the results in Table 2, the estimated capacity values based on this method are mostly lower than the average pre-breakdown flow rates, the PLM, the SFI, and the Van Aerde capacities, with the exception of Site 8, which provided a higher estimate for capacity. It should be noted that at Site 8, the HCM6 Weibull curve is very close to the PLM Weibull curve (Figure 2*e*), whereas at the remaining sites the curves are significantly different. The difference in the Weibull curves and resulting capacities between the two methods can be attributed to the fact that the HCM6 method fits the Weibull distribution to the probability of breakdown in each bin and not to the cumulative probability of breakdown that the PLM (and SFI) uses.

It appears that the HCM6 probability function is greatly affected by the probability ratio at higher volume bins, as it was not possible to fit the Weibull distribution at Site 7, where the probability ratios were very low. To evaluate this issue, a sensitivity analysis was performed, in which the HCM6 method was reapplied to site 1 while considering a minor adjustment to the number of breakdown observations. Table 3 shows the original and modified number of observations and breakdowns in each bin along with the corresponding probabilities (ratios). To apply the adjustment, one breakdown observation was added to bin 19 (1,900–2,000 vphpl). The Weibull distribution was then fitted to the modified data and the estimated capacity was computed using Equation 6. The same procedure was performed on the PLM to evaluate its sensitivity to the input data. Therefore, one observation was added to the same interval and the Weibull distribution was again fitted to the modified data. The modified and the original fitted curves for both PLM and HCM6 methods are shown in Figure 3.

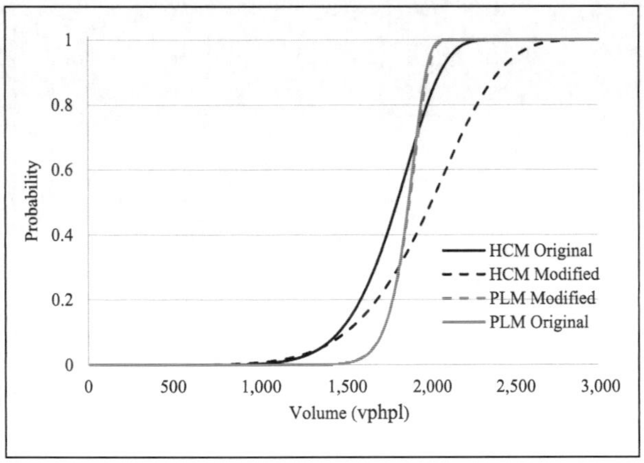

Figure 3. Original and modified Weibull distributions based on HCM6 and PLM for Site 1 (I-35 SB South of 67th Street).

The HCM6 capacities were 1,608 vphpl and 1,538 vphpl for the original and modified data, respectively; a 70 vphpl difference (approximately 5%). According to Figure 3, the fitted HCM6 Weibull distribution is shifted noticeably to the right by applying the minor adjustment. The comparison reveals that a single breakdown observation within a bin at high volume will result in significant variation of the fitted distribution and the estimated capacity. The difference magnitude has inverse correlation to the number of observations within each bin. In other words, a lower number of observations within each bin will result in a more significant difference in capacity. On the other hand, the fitted Weibull curve from the PLM did not seem to change from its original shape (dashed and solid gray lines in Figure 3 overlap). This suggests that a single breakdown observation has an insignificant effect on the fitted curve and the corresponding estimated capacity. As such, it can be concluded that the PLM takes into account the overall distribution rather than single observations, whereas the HCM6 method is very sensitive to the number of breakdown observations within each bin.

Conclusions and Recommendations

This paper investigated the freeway capacity estimation by performing an overall comparison of previously developed methods. Two previously investigated methods, namely the PLM and the Van Aerde method, as

Table 4. Advantages and Disadvantages of Capacity Estimation Methods

Method	Advantages	Disadvantages
Van Aerde	• Easy application and fitting • Breakdown identification is not required • Capacity is not influenced by single observations • Does not require intensive computations	• Not quite effective on segments that do not capture the entire spectrum of congested/uncongested conditions • Does not consider breakdown observation for estimating capacity
PLM	• Estimates capacity based on a probability distribution function • Based on breakdown observations • Capacity influenced not by a single breakdown observation but by the entire breakdown data set	• Requires intensive data computation • Requires breakdown identification • Not applicable on sites with low number of breakdown observations
SFI	• Yields the optimum flow rate that corresponds to the volume with highest survival probability • Based on breakdown observations • The optimum flow is influenced not by a single breakdown observation but by the entire breakdown data set	• Requires intensive data computation • Requires breakdown identification • Estimated capacities are usually lower compared with other methods but higher than the HCM6 method
HCM6	• Estimates capacity with a single-point estimate based on the breakdown probability distribution • Based on breakdown observations • Considers the ratio (probability) of breakdown within various volume intervals	• Requires intensive data computation • Requires breakdown identification • Capacity is very sensitive to individual breakdown observation rather than to the whole breakdown data set • Not applicable on sites with a low number of breakdown observations • The estimated value is usually lower compared with other methods

well as two newly developed models, namely the HCM6 and the SFI, were evaluated using data at multiple merge junctions along Kansas City urban freeways. According to the results, the PLM Weibull-fitted model provided on average the highest capacity estimates, followed by the Van Aerde model capacities. The Van Aerde capacities were also found to be closer to the average pre-breakdown flow rates. The estimated capacities based on the SFI method seemed to vary significantly compared to the average pre-breakdown flow rates. In three out of the five sites, the SFI yielded higher estimates for capacity in comparison to the average pre-breakdown flow rates and the Van Aerde capacities. The SFI also provided lower capacity estimates compared to the PLM. Lastly, the HCM6 method resulted in the lowest capacity estimates for most of the sites except one (Site 8). Both PLM and HCM6 are highly dependent on the number of breakdown observations, and the actual number of observations in each bin seems to affect the HCM6 model more. The advantages and the disadvantages of each method are summarized in Table 4.

It is concluded that the Van Aerde method, regardless of its convenient application, could not fully represent the freeway capacity because it does not consider the breakdown events or the breakdown flow distribution. Additionally, the process of calibrating the estimated values for Van Aerde heavily depends on the overall speed-flow scatter diagram. In these cases, as it is difficult to observe the actual speed drop location and the corresponding capacity (Figure 1*f*), a single estimate for capacity may not be appropriate. However, the HCM6 method showed a great deal of variability among the different sites and resulted in the lowest capacity estimates. The HCM6 method is considerably affected by the breakdown ratio within each volume bin, and more so at higher volume bins. As described in the section on results and discussion, a single breakdown observation can significantly shift the fitted distribution and the corresponding estimate for capacity. Additionally, the HCM6 method failed to produce an acceptable distribution and capacity estimate at one site (Site 7) because of the low breakdown probability ratio. The PLM Weibull-fitted method provided the highest capacity estimates and more consistent results across all sites. The PLM was able to take into account the overall data set and was not found to be sensitive to individual breakdown observations. Nevertheless, the PLM also failed to produce an acceptable distribution and capacity estimate at one site because of a low number of breakdown events. Lastly, the SFI method closely follows the PLM results, but it was found to provide lower (more conservative) capacity estimates.

The comparison of the different capacity estimation methods investigated in this paper suggests that breakdown-based probabilistic methods are generally preferred. These types of models have the ability to capture the stochastic nature of capacity; however, they require a significant amount of data and effort for identifying the breakdown events. The PLM and the resulting SFI method showed greater consistency compared to the HCM6 method. It appears that the difference between the HCM6 and the PLM (and SFI) is attributed to the fact that the HCM6 method fits the Weibull distribution to the probability of breakdown in each bin and not to the cumulative probability of breakdown that the PLM (and SFI) uses. Nonetheless, further research should look into the HCM6 method and under which conditions it can produce consistent results. The SFI also showed similar results with the average pre-breakdown flow rate. Finally, selecting a single capacity value from a distribution by considering an acceptable probability (such as in the HCM6 method) or through another mechanism (such as in the SFI method) significantly improves the usefulness and applicability of that method.

Acknowledgments

The authors would like to thank KC Scout for providing the traffic data for this research.

Author Contributions

The authors confirm contribution to the paper as follows: study conception and design: Alexandra Kondyli; data collection: Mohamadamin Asgharzadeh; analysis and interpretation of results: Alexandra Kondyli, Mohamadamin Asgharzadeh; draft manuscript preparation: Mohamadamin Asgharzadeh. All authors reviewed the results and approved the final version of the manuscript.

References

1. Kondyli, A., L. Elefteriadou, W. Brilon, F. L. Hall, B. Persaud, and S. Washburn. Development and Evaluation of Methods for Constructing Breakdown Probability Models. *Journal of Transportation Engineering*, Vol. 139, No. 9, 2013, pp. 931–940.
2. Transportation Research Board, Highway Capacity Manual. *A Guide for Multimodal Mobility Analysis*, 6th ed. Transportation Research Board, Washington, D.C., 2016.
3. Greenshields, B., W. Channing, and H. Miller. A study of Traffic Capacity. *Proc., Highway Research Board No. 1935*, National Research Council (USA), Highway Research Board, 1935.
4. Van Aerde, M. Single Regime Speed-Flow-Density Relationship for Congested and Uncongested Highways. Presented at 74th Annual Meeting of the Transportation Research Board, Washington, D.C., 1995.
5. Lorenz, M., and L. Elefteriadou. A Probabilistic Approach to Defining Freeway Capacity and Breakdown. *Proc., 4th International Symposium on Highway Capacity*, Transportation Research Board, Maui Hawaii, 2000, pp. 84–95.

6. Elefteriadou, L., R. P. Roess, and W. R. McShane. Probabilistic Nature of Breakdown at Freeway Merge Junctions. *Transportation Research Record: Journal of the Transportation Research Board*, 1995. 1484: 80–89.
7. Hall, F. L., and K. Agyemang-Duah. Freeway Capacity Drop and the Definition of Capacity. *Transportation Research Record: Journal of the Transportation Research Board*, 1991. 1320: 91–98.
8. Minderhoud, M. M., H. Botma, and P. H. Bovy. Roadway Capacity Estimation Methods Explained and Assessed. *Transportation Research Record: Journal of the Transportation Research Board*, 1997. 1572: 59–67.
9. Brilon, W., and H. Zurlinden. Breakdown Probability and Traffic Efficiency as Design Criteria for Freeways. *Forschung Strassenbau und Strassenverkehrstechnik*, Vol. 870, 2003.
10. Brilon, W., J. Geistefeldt, and M. Regler. Reliability of Freeway Traffic Flow: A Stochastic Concept of Capacity. *Proc., 16th International Symposium on Transportation and Traffic Theory*, College Park, Maryland, 2005.
11. Brilon, W. Traffic Flow Analysis Beyond Traditional Methods. *Proc., 4th International Symposium on Highway Capacity*, Transportation Research Board, Maui Hawaii, 2000, pp. 26–41.
12. Shojaat, S., J. Geistefeldt, S. A. Parr, L. Escobar, and B. Wolshon. Applying the Sustained Flow Index to Estimate Freeway Capacity. Presented at 96th Annual Meeting of the Transportation Research Board, Washington, D.C., 2017.
13. Lorenz, M., and L. Elefteriadou. Defining Freeway Capacity as Function of Breakdown Probability. *Transportation Research Record: Journal of the Transportation Research Board*, 2001. 1776: 43–51.
14. Kühne, R., R. Mahnke, and J. Hinkel. Modelling the Effects of Corridor Control Systems on Road Capacity. *Proc., 5th International Symposium on Highway Capacity and Quality of Service*, Yokohama, Japan, 2006.
15. Modi, V., A. Kondyli, S. S. Washburn, and D. S. McLeod. Freeway Capacity Estimation Method for Planning Applications. *Journal of Transportation Engineering*, Vol. 140, No. 9, 2014, p. 05014004.
16. Kennedy, J. Particle Swarm Optimization. In *Encyclopedia of Machine Learning*, Springer, 2010, pp. 760–766.
17. Kaplan, E. L., and P. Meier. Nonparametric Estimation from Incomplete Observations. *Journal of American Statistical Association*, Vol. 53, No. 282, 1958, pp. 457–481.

The Standing Committee on Highway Capacity and Quality of Service (AHB40) peer-reviewed this paper (18-03976).

Article

TRR
JOURNAL OF THE TRANSPORTATION RESEARCH BOARD

Transportation Research Record
2018, Vol. 2672(15) 85–92

DOI: 10.1177/0361198118782253
journals.sagepub.com/home/trr
SAGE

Modeling the Impacts of Traffic Flow Arrival Profiles on Ramp Metering Queues

Guangchuan Yang[1], Rui Yue[1], Zong Tian[1], and Hao Xu[1]

Abstract

An adequate queue storage length is critical for a metered on-ramp to prevent ramp queue spillback to the upstream signalized intersection. Previous research on queue length estimation or queue storage length design at metered ramps has not taken into account the potential impact of various on-ramp traffic flow arrival profiles on ramp queue lengths. This paper depicts the traffic flow arrival profiles and queue generation processes at three different metered ramp categories. Based on a large number of microscopic simulation runs, it is found that, under a given demand-to-capacity scenario, the queue at a metered ramp with two on-ramp feeding movements is more likely to be cleared in a cycle than at a metered ramp with three on-ramp feeding movements. Also, the platoon dispersion effect significantly reduces the ramp queue length, and hence the queue storage needs at a metered ramp. In addition, this paper reveals that ramp queue length tends to increase linearly with upstream signal cycle length. The design of queue storage length for a metered on-ramp hence needs to fully consider the various ramp configurations and upstream signal timing settings.

Ramp metering uses traffic signals at freeway entrance ramps to regulate the rate at which vehicles enter the freeway. It temporarily stores the entering traffic on the ramps to alleviate congestion on freeways using the "access rate reduction" technique (*1*). This "access rate reduction" technique ensures the total traffic entering the freeway section remains below the operational capacity of the freeway section. Another objective of ramp metering is to break up the platoons of vehicles that have been released from upstream arterial traffic signals entering the freeways. Under normal conditions, the freeway mainline can accommodate merging vehicles one or two at a time even when operating near capacity; however, when groups of vehicles attempt to force their way into freeway traffic, turbulence and shockwaves will be created at the merging areas, causing the mainline flow to break down (*2*, *3*). Therefore, ramp metering could reduce the turbulence that may cause congestion around the merging area and help to prevent freeway capacity drop. In addition, vehicles individually released onto the freeway mainline will lead to a reduction in rear-end and sideswipe type of crashes, and can promote a smoother and safer merging maneuver (*4*). A cross-sectional study performed by the Minnesota Department of Transportation turned off more than 400 ramp meters in the Minneapolis-St. Paul area to compare the traffic operations at the existing freeway system. The results showed that the freeway throughput decreased by 9%, travel time increased by 22%, speeds dropped by 7%, and crashes increased by 26% (*5*).

However, regulation of on-ramp traffic will result in a queue formed in front of the ramp meter. In current practice, the majority of metered ramps are retrofitted to existing unmetered ramps; under this condition, the queue storage space is usually insufficient. When adequate queue storage is not provided, ramp queue may spillback to the upstream signalized intersection, which would significantly affect the operations of the arterial (*6*). Therefore, an accurate estimation of the required queue storage length is critical to the operation of ramp metering.

A number of studies have been performed for estimating queue length at metered ramps. Liu et al. (*7*) found when the counting errors of ramp queue and passage detectors are less than 2%, queue length can be estimated on the basis of flow conservation; while for ramps which have significant counting error (i.e., larger than 2%), the Kalman-Filter method provided more accurate estimation of queue lengths. Vigos et al. (*8*) employed a Kalman-Filter estimator with real-time traffic flow and occupancy detector data for metered ramp queue length estimation. Based on microscopic simulation, the authors pointed out that Klaman-Filter estimator could provide reasonable and reliable estimations of queued vehicles, and the

[1]Department of Civil and Environmental Engineering, University of Nevada, Reno, NV

Corresponding Author:
Address correspondence to Guangchuan Yang: yanggc1986@gmail.com

fundamental issues that may affect ramp queue length include: the effects of loop density, update period, downstream signal cycle, vehicle length, and link length. Similarly, Wu et al. (*9*) found that when direct measurement is not an option, the Kalman-Filter method provided a better prediction of real-time queue lengths, while field-observed queue data should be collected for careful calibration and operator verification. Lee et al. (*10*) designed a Kalman-Flter-based framework for estimating ramp queue length. The flow conservation model was employed to estimate the system state, then the time occupancy measurements from the mid-link and link entrance detectors were used for continuously correcting the system state. Sanchez et al. (*11*) proposed a method using magnetic sensors with vehicle real-time re-identification technique for queue length estimation at metered on-ramps. Results show that this method correlated well with field-observed data when a ramp was uncongested, but it underperformed during saturated ramp conditions. Yang et al. (*12*, *13*) employed both an analytical approach and field-observed data to illustrate the difference between the platoon traffic flow arrival pattern at metered arterial-to-freeway ramp and the random traffic flow arrival pattern at metered freeway-to-freeway connector. Based on a mesoscopic queue length simulation method, the authors investigated the impacts of platoon and random traffic flow arrival patterns on queue length at the two different metered on-ramps; their results indicate that under similar demand-to-capacity conditions, the platoon arrivals will result in longer queues.

The existing models presented so far show that there is still no clear insight into how on-ramp traffic flow arrival profile is affected by ramp geometric features and upstream signal setting, and there is no dedicated study that aims to reveal the impacts of various on-ramp traffic flow arrival profiles on the queuing process at metered ramps. In current practice, a ramp meter is usually located immediately downstream of a signalized intersection. As a result of the cyclic operation of the upstream signal, for each on-ramp feeding movement, traffic flow tends to enter the ramp first in a platoon discharged from the green signal, then in a non-platoon pattern which can be visualized as a uniform arrival rate. This indicates that traffic flow arrival rate is non-uniform within each signal cycle. In general, the platoon arrival traffic in each cycle will generate a longer queue length while the non-uniform arrival traffic tends to result in a short queue length. Also, due to the stochastic nature of traffic flow, the cyclic on-ramp traffic volume tends to fluctuate which usually follows a certain distribution pattern. Therefore, considering the different ramp geometric configurations (i.e., upstream signal release pattern) and traffic demand levels, the on-ramp traffic will arrive at the ramp meter under various flow profiles.

Several studies have been conducted with the aim of incorporating the impacts of traffic flow arrival pattern on traffic performance analysis at signalized intersections. Akcelik and Rouphail (*14*) indicated that, for a paired signalized intersection system, traffic flow is released in platoon pulses at a virtually uniform rate each cycle; in this case, the application of the random arrivals model to platooned arrivals may not be able to accurately estimate queue lengths and delays. Lin et al. (*15*) pointed out that the actual queue discharge pattern at signalized intersections usually does not display a steady saturation rate, hence the use of a saturation flow rate may mislead the planning and operational decisions. Geroliminis and Skabardonis (*16*) proposed a Markov decision process model to predict the platoon arrival profiles and queue length at signalized links. Kinematic wave theory was employed to capture traffic dynamics. Bonneson et al. (*17*) indicated that an accurate prediction of the aggregated arrival flow profile at the downstream of a signalized link should separately account for the effects of platoon dispersion or platoon decay. However, in reality, the geometric configuration and queuing process of a ramp metering system are significantly different from a paired signalized intersection system along a signalized arterial. Though the traffic flow arrival profile at a metered ramp is similar to the traffic flow at an approach to a signalized intersection, which can be visualized as platoon arrival pulses with a uniform arrival during each phase, they have different downstream queue discharge patterns. For a paired signalized intersection system, the queued vehicle will be discharged at the approach's saturation flow rate during the green period. In comparison, at metered ramps, the queued vehicles will be discharged at the ramp meter's average metering rate, which is usually much lower than its saturation flow rate. Hence, the real-time queue length profile and the maximum queue length during a cycle of a metered ramp will differ from a paired signalized intersection system. This indicates that previous research findings about queue length at signalized intersections will not be applicable for metered ramps. Therefore, to provide more accurate estimation of the queue storage length needs at metered on-ramps, the average on-ramp flow rate, the cycle-by-cycle demand fluctuations, and the effects of degree of traffic platooning should all be incorporated in the traffic performance analysis to capture more accurately the actual on-ramp traffic flow arrival profiles.

In summary, previous studies demonstrated that traffic flow arrival profile is a critical factor for accurately predicting queue lengths. So far, no dedicated study has attempted to investigate the potential impacts of traffic flow arrival profiles on queue storage needs at various metered ramp configurations. Therefore, this paper aims to present an analytical method for modeling the traffic flow arrival profiles at various metered on-ramps, and to investigate the impacts of different on-ramp flow arrival profiles on ramp queue lengths under various demand-to-capacity ratio scenarios based on microscopic simulation.

Modeling Traffic Flow Arrival Profiles

Three Metered Ramp Configurations

Based on the field investigation, this paper defines three typical metered ramp configurations with different on-ramp traffic flow arrival profiles. These ramps are categorized

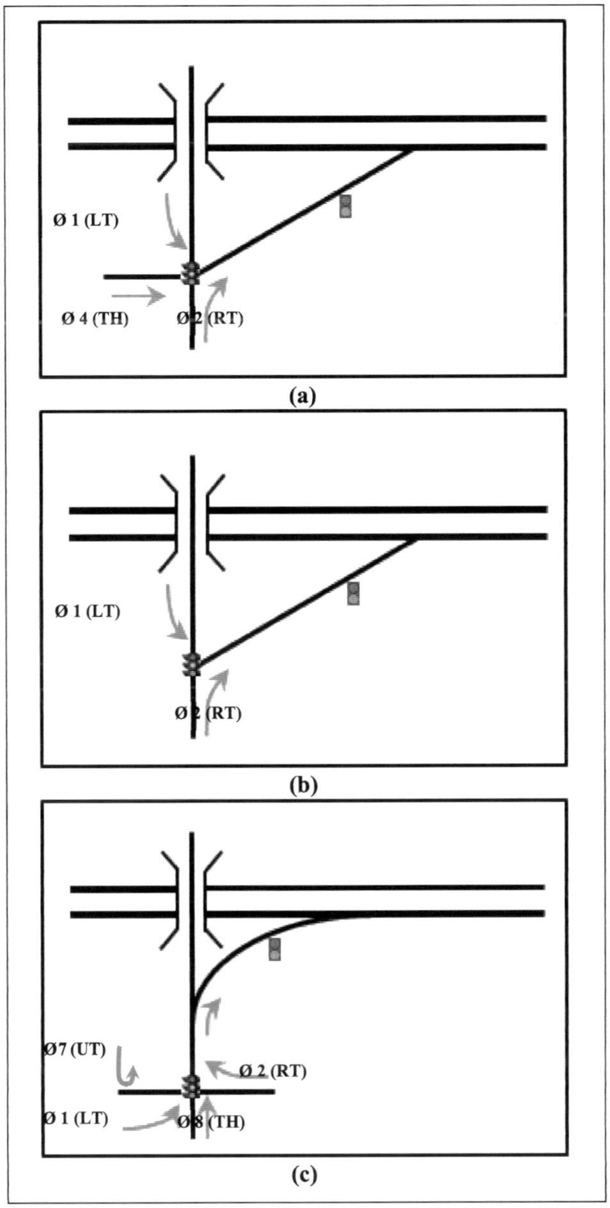

Figure 1. Geometric configuration of three typical metered ramps: (*a*) Category 1 metered ramp, (*b*) Category 2 metered ramp, and (*c*) Category 3 metered ramp.

mainly based on the number of on-ramp feeding movements.

Figure 1*a* shows a conventional diamond interchange with three movements feeding to the metered ramp, which is defined as Category 1. The frontage road northbound through movement, and the arterial left turn and right turn movements, are controlled by the upstream signal, usually under a three-phase signal control strategy.

Figure 1*b* illustrates another ramp configuration common in the United States: a tight diamond interchange with two on-ramp feeding movements, which is defined as Category 2. In some cases, the off-ramp vehicles can re-enter the freeway. However, field observation showed volume of this movement is very limited, usually less than 1% of the total on-ramp feeding volume, thus was ignored in this study. In this case, the arterial left turn and right turn on-ramp feeding movements contribute to the ramp queue. For this ramp configuration, the upstream signal usually has three phases, while only two phases are on-ramp feeding phases.

Figure 1*c* describes the third common ramp configuration. The ramp arrival flow pattern of this ramp type differs from the other two categories of ramps. Typically there are three movements feeding to the ramp under a four-phase upstream signal control strategy (four feeding movements can exist when there are arterial vehicles making a U-turn to enter the freeway on-ramp). Traffic flow diverges at the ramp entrance: part of the traffic from the upstream intersection enters the metered ramp and the remainder stays on the arterial road. The key differences between Category 3 and Category 1 include: (a) the upstream intersection departures will affect the upstream signal timing and thus result in different on-ramp arrival flow patterns; this is particularly significant for actuated signals; (b) since a portion of upstream departure flow diverges to the arterial, the platoon from the upstream intersection is partially dispersed. The queue length estimation model should therefore take into account the difference of on-ramp platoons.

Description of On-Ramp Traffic Flow Arrival Profiles

Due to the existence of the upstream signals, the on-ramp flow arrives at the ramp meter with unique patterns, which are related to the type of control and the signal timing setting at the upstream signals. The actual on-ramp arrival profiles are mainly influenced by the upstream signal phasing and timing, number of feeding movements, and average arrival rate of each feeding movement. Based on the analysis of field collected traffic flow data and using analytical method, the on-ramp flow arrival profile for each on-ramp category is depicted, as illustrated in Figure 2.

In Figure 2*a*, the arrival profile is specifically for a metered ramp with three feeding movements at the upstream intersection. Traffic from each feeding movement arrives at the ramp meter in two flow regimes: the regime with saturated upstream intersection discharging flow (platoon arrival) and the regime with traffic flow equal to the arrival flow of the upstream intersection (non-platoon arrival). For example, during the frontage road through movement (*TH*, or phase $\phi 4$), the first portion of flow occurs when the frontage road through movement discharges at its saturation flow rate S_{TH}. After the vehicle queue of the through movement is cleared, the flow rate reduces to its average arrival rate A_{TH}. The primary difference between Category 2 ramp and Category 1

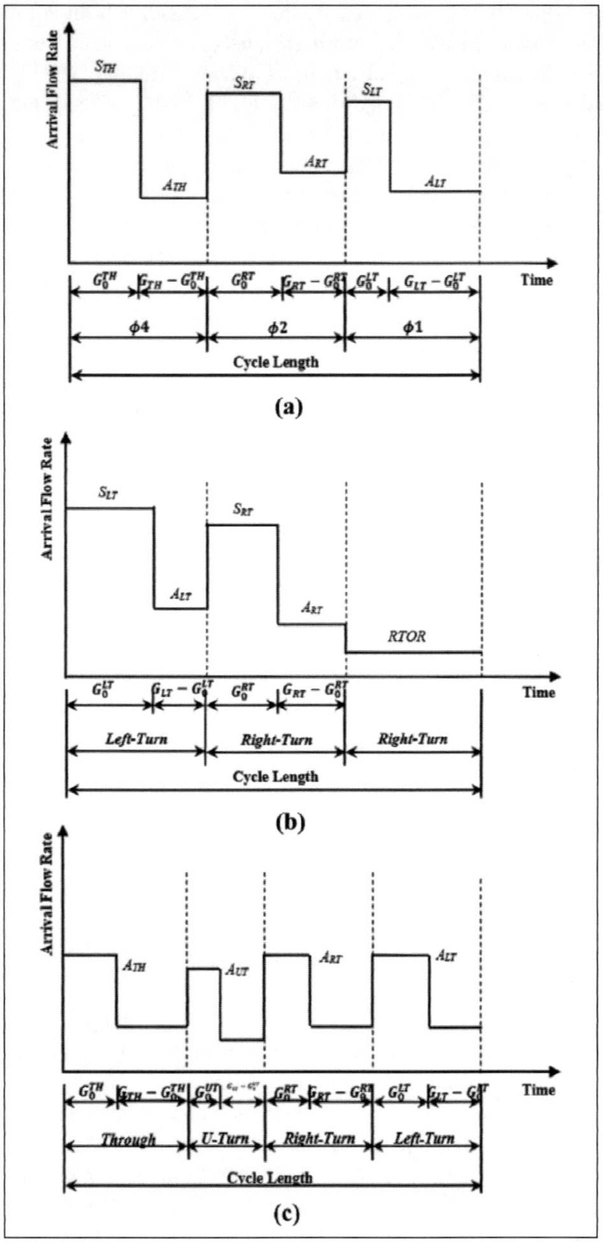

Figure 2. On-ramp flow arrival profile at: (*a*) Category 1 metered ramp, (*b*) Category 2 metered ramp, and (*c*) Category 3 metered ramp.

ramp is that during phase $\phi 4$ there is no direct on-ramp feeding movement. There will be a small portion of right-turn-on-red vehicles entering the ramp, thus the on-ramp arrival rate during this phase is much lower than other phases, as illustrated in Figure 2*b*. Category 3 ramp has a similar upstream traffic flow discharging pattern; however, due to the platoon dispersion effects, traffic flow usually arrives at the ramp meter at a rate that is lower than the saturation flow rate.

Queue Generation at Metered Ramps

A metered ramp is a standard queuing system; for each analysis time interval, vehicle queues will be formed when the vehicle arrival rate during this time interval exceeds the ramp metering rate. Ramp queues may exist temporally due to a short-term surge of traffic arrival even during under-saturated conditions (e.g., a platoon arrival that released from upstream signal) or it can be prolonged due to over-saturation over an extended period of time. The queuing process at each metered ramp category is demonstrated in Figure 3 (under-saturated condition) and Figure 4 (over-saturation condition). The formed queues can be estimated deterministically using queuing theory with certain assumptions (*18*, *19*). Field investigations show that even though sophisticated ramp metering systems can adjust ramp metering rates on the basis of traffic conditions, metering rates are usually uniform over a period longer than a signal cycle. Therefore, for simplicity in demonstrating the queuing process, this paper assumed a fixed ramp metering rate; thus the departure flow was uniform. In consistency with Figure 2, for Category 1 and 2 ramps, and for each on-ramp feeding movement, the platoon arrival rate equals its saturation flow rate; after that, on-ramp traffic is assumed to arrive at the ramp at the average arrival rate. For the Category 3 ramp, since the vehicle platoons tend to be dispersed before entering the ramp, traffic flow arrives at the ramp meter in a relatively random pattern where the arrival rate is less than the saturation flow rate.

Queue Length Simulation

To investigate the impacts of different on-ramp traffic flow arrival profiles on ramp queue lengths, this research employed the PTV VISSIM microscopic simulation software for simulating the queue lengths of each metered ramp category. To eliminate the potential uncertainties that may be involved in the simulation (such as the lane-changing behavior and un-balance distribution of traffic flow when there are multiple lanes), in this study the default number of lanes is one lane for each on-ramp feeding movement, and both on-ramp traffic demand and ramp metering rate are standardized as vehicle-per-hour-per-lane (vphpl). Various scenarios were created for different combinations of on-ramp demands and metering rates. In this paper, only under-saturation scenarios are considered, since the design of ramp meter facilities needs to avoid an over-saturation situation. The designed demand-to-capacity ratios range from approximately 0.3 to 1.0, as listed in Table 1. In this study, the queue length reporting interval was set at 90 seconds, which equals the cycle length of the upstream signal. Hence, each simulation run will generate 40 queue lengths. Since the queue storage length design for a metered ramp needs to consider the most challenging situation, this research selected the maximum queue length among the 40 queue lengths for further data

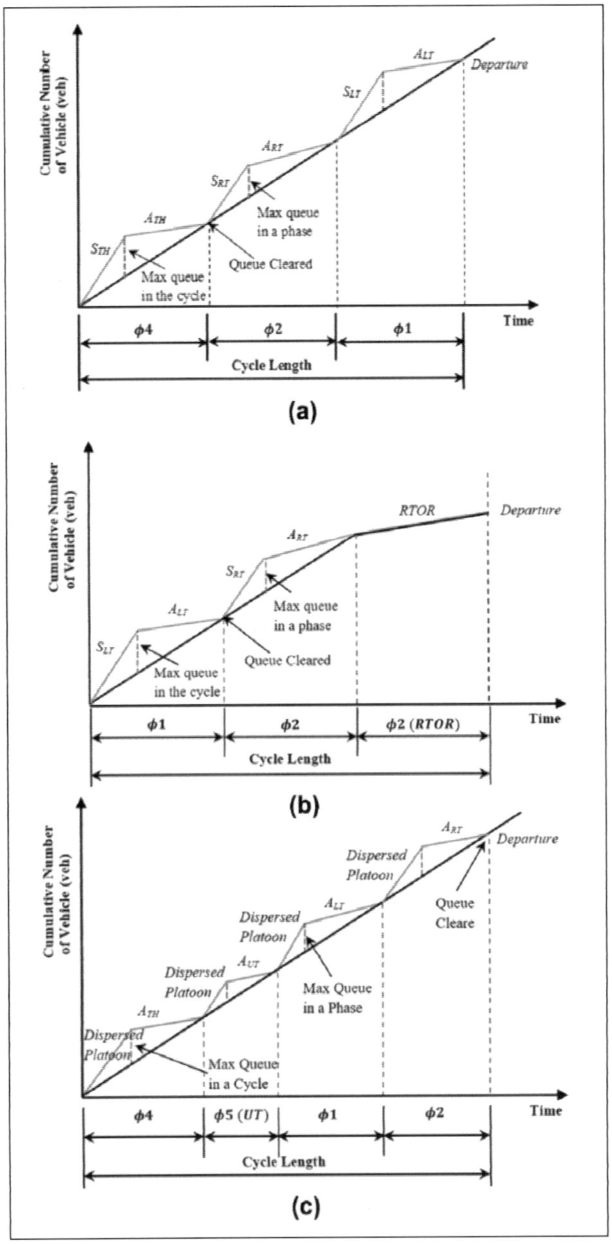

Figure 3. Queue formed by under-saturated arrival traffic at: (*a*) Category 1 metered ramp, (*b*) Category 2 metered ramp, and (*c*) Category 3 metered ramp.

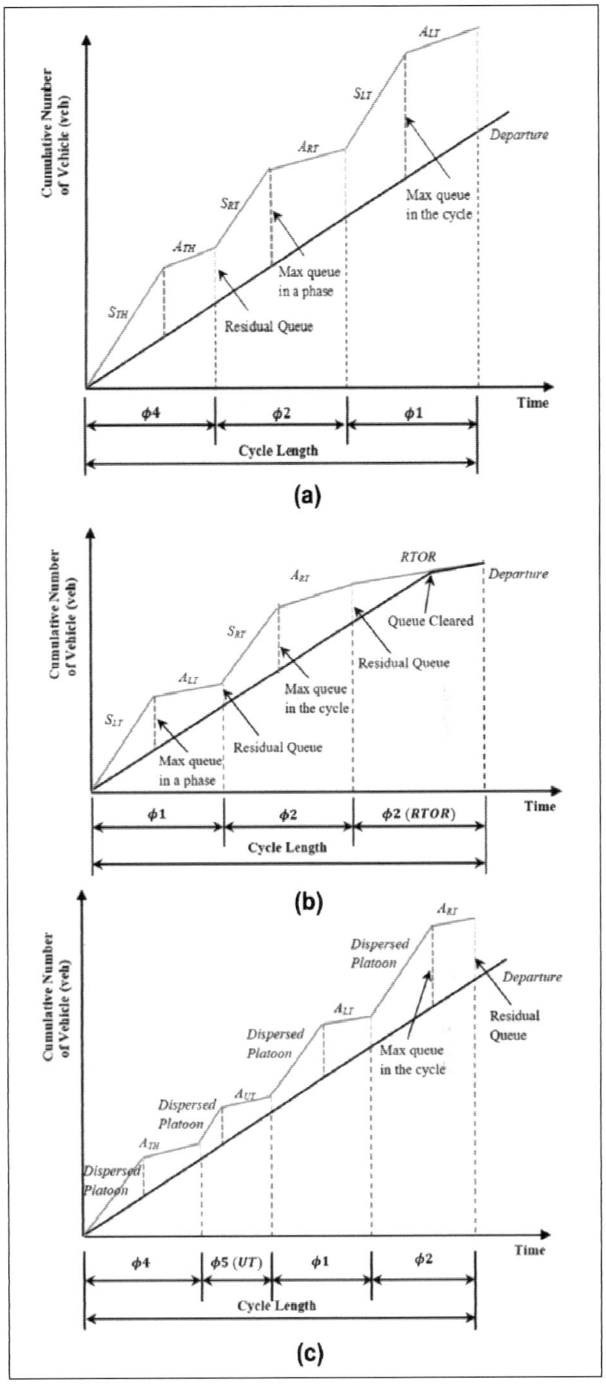

Figure 4. Queue formed by over-saturated arrival traffic at: (*a*) Category 1 metered ramp, (*b*) Category 2 metered ramp, and (*c*) Category 3 metered ramp.

analysis. For each scenario, 10 simulations were performed to obtain the mean of these maximum queue lengths.

Impact of Demand-to-Capacity Ratio on Ramp Queue Length

Based on the deterministic queuing theory, the primary factor that affects ramp queue length is the demand-to-capacity ratio. With the aforementioned simulation scenarios, microscopic simulations were conducted for the three metered on-ramp categories. Based on a large number of simulation runs, the maximum queue length of each scenario is extracted, as illustrated in Figure 5.

For each ramp category, a regression equation was developed to calculate the correlation between queue length and demand-to-capacity ratio, as depicted in Figure 6.

Table 1. Summary of Queue Length Simulation Scenarios

Ramp metering cycle length (sec)	Metering rate (vphpl)	On-ramp demand (vphpl)	Demand-to-capacity ratio
12	300	100–300	0.33–1.0
11	327	100–300	0.31–0.92
10	360	100–350	0.28–0.97
9	400	150–400	0.38–1.0
8	450	150–450	0.33–1.0
7	514	150–500	0.29–0.97
6	600	200–600	0.33–1.0
5	720	250–700	0.35–0.97

Note: On-ramp demand increases from the low boundary to the high boundary in 50 vphpl intervals. vphpl = vehicle-per-hour-per-lane.

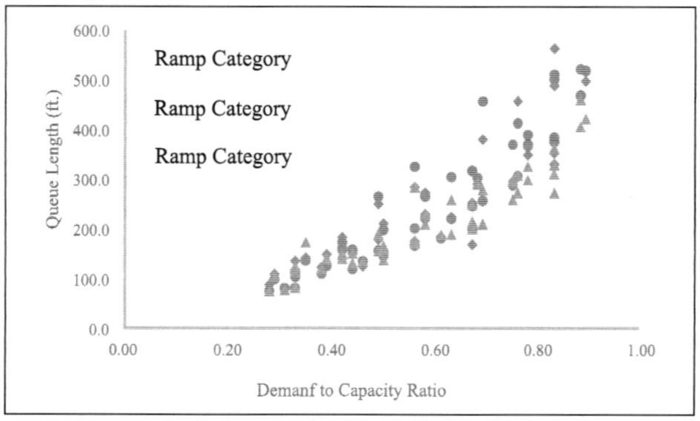

Figure 5. Queue length scatter plots of the three ramp categories.

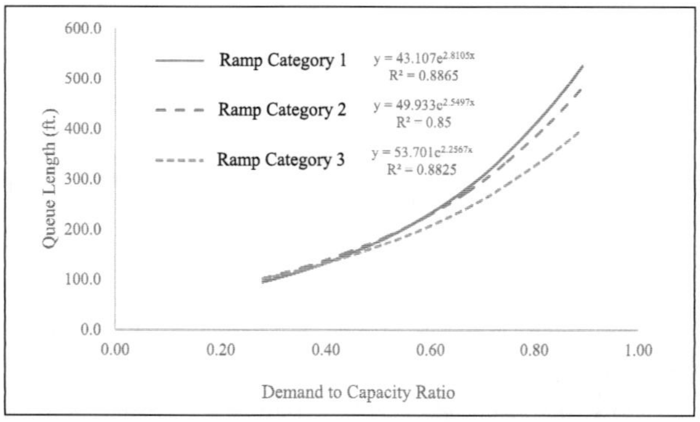

Figure 6. Queue length versus demand-to-capacity ratio profiles of the three ramp categories.

Results show that, at aggregate level, when demand-to-capacity ratio is lower than 0.5, there is no significant difference between the queue lengths of the three metered ramp categories. With the increase of demand-to-capacity ratio, queue lengths of the Category 1 and 2 metered ramps increase in a more dramatic pattern in comparison with the Category 3 metered ramp. This indicates that under similar demand-to-capacity scenarios, the platoon dispersion effect at a metered ramp tends to reduce the queue storage needs. In comparison with the Category 1 metered ramp, shorter queue lengths were found at the Category 2 metered ramp, particularly when demand-to-capacity ratio is higher than 0.8. A reason for this phenomenon is that there are three on-ramp feeding movements at the Category 1 metered ramp, so during each cycle there are three vehicle platoons entering the ramp. For the Category 2 metered ramp, however, only two vehicle platoons enter the ramp, and during the third phase the right-turn-on-red traffic rate is usually lower than the ramp metering rate, hence the queue is more likely to be cleared than at the Category 1 metered ramp.

Impacts of Upstream Signal Cycle Length on Ramp Queue Length

Another important factor that may affect ramp queue length is the upstream signal setting, since a change of the upstream signal cycle length would produce a different on-ramp flow arrival profile. To investigate the impacts of different upstream signal cycle length on ramp queue length, this paper used the Category 1 metered ramp for queue length simulation under various cycle length scenarios. Based on the field investigation, the default cycle length was set as 90 seconds, and the testing cycle lengths ranged from 60 to 120 seconds with 10 seconds increase interval. The on-ramp demand and metering rate were set as 500 and 600 vphpl, respectively, which yields a demand-to-capacity ratio of 0.83. As mentioned before, the maximum queue length was used for data analysis and for each scenario, 10 simulations were performed to obtain the mean of the maximum queue lengths. The simulated maximum queue lengths under the predetermined cycle length scenarios are plotted in Figure 7.

Results show that the maximum queue length tends to increase linearly with cycle length. This is mainly because, given a longer cycle length, for each on-ramp feeding movement there will be a longer platoon arrival period; with these longer vehicle platoons, the ramp queue length tends to be extended within a short period. In this regard, the design of metered ramp queue storage length also needs to take into account the potential impacts of upstream signal cycle length on the on-ramp traffic flow arrival profile.

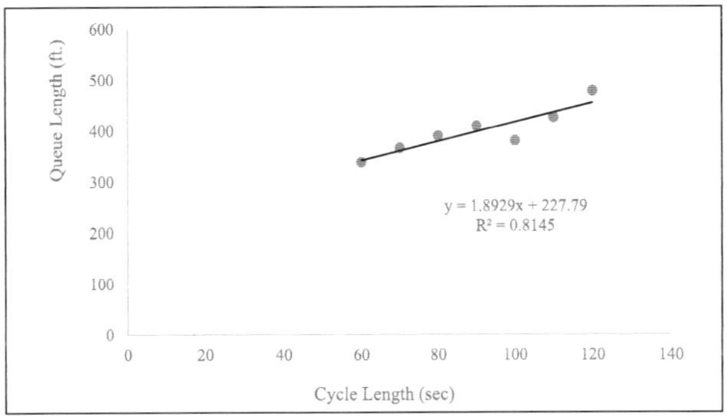

Figure 7. Queue length scatter plots for various cycle length scenarios.

Conclusion

This paper reveals that the traffic flow arrival profile is affected by both ramp geometric features and upstream signal setting. Simulation results indicate that the traffic flow arrival profile has significant effects on the queue length generation at metered ramps. It was found that under a given demand-to-capacity scenario, the platoon dispersion effect significantly reduced the ramp queue length; therefore, for the Category 3 metered ramp where a vehicle platoon discharged from the upstream signal dispersed before entering the ramp, a shorter ramp queue length storage space may satisfy the on-ramp demand. Also, it was found that the queue at a metered ramp with two on-ramp feeding movements (i.e., Category 2 metered ramp) is more likely to be cleared in a cycle than at a metered ramp with three on-ramp feeding movements (i.e., Category 1 metered ramp). This indicates that, under the same traffic conditions, Category 2 metered ramps require less queue storage space than Category 1 metered ramps. This paper also found that ramp queue length tends to increase linearly with upstream signal cycle length. Hence, the design of metered ramp queue storage length needs to fully consider the various ramp configurations and upstream signal settings.

The research findings presented in this paper were based on the assumptions of fixed upstream signal cycle length and metering rate; future works need to further investigate the impacts of actuated upstream signal control and traffic responsive ramp metering strategies on ramp queue length. This will help to develop an integrated upstream signal and ramp metering control strategy. Also, the simulated queue lengths may not exactly represent the real-world condition; to make the queue length simulation results more feasible for real-world implementation, future works need to further validate the simulated queue lengths using field collected data at different metered ramp configurations and under various demand-to-capacity conditions.

Acknowledgments

This study was sponsored by the California Department of Transportation (Caltrans), grant number 65A0486. The authors thank Zhongren Wang of Caltrans Headquarters, and Arafat Khan and Yue Zhao of the University of Nevada, Reno, for their constructive comments.

Author Contributions

The authors confirm contribution to the paper as follows: study conception and design: Zong Tian, Guangchuan Yang; Hao Xu; data collection: Guangchuan Yang; Rui Yue; analysis and interpretation of results: Guangchuan Yang, Rui Yue; draft manuscript preparation: Guangchuan Yang. All authors reviewed the results and approved the final version of the manuscript.

References

1. Papageorgiou, M., and A. Kotsialos. Freeway Ramp Metering: An Overview. *IEEE Transactions on Intelligent Transportation Systems*, Vol. 3, No. 4, 2002, pp. 271–281.
2. Cassidy, M. J., and J. Rudjanakanoknad. Increasing the Capacity of an Isolated Merge by Metering Its On-Ramp. *Transportation Research Part B: Methodological*, Vol. 39, No. 10, 2005, pp. 896–913.
3. Zhang, L., and D. Levinson. Ramp Metering and Freeway Bottleneck Capacity. *Transportation Research Part A: Policy and Practice*, Vol. 44, No. 4, 2010, pp. 218–235.
4. Lee, C., B. Hellina, and K. Ozbay. Quantifying Effects of Ramp Metering on Freeway Safety. *Accident Analysis and Prevention*, Vol. 38, No. 2, 2006, pp. 279–288.
5. Levinson, D., and L. Zhang. Ramp Meters on Trial: Evidence from the Twin Cities Metering Holiday. *Transportation Research Part A: Policy and Practice*, Vol. 40, No. 10, 2006, pp. 810–828.
6. Tian, Z., C. Messer, and K. Balke. Modeling the Impact of Ramp Metering Queues on Diamond Interchange Operations. *Transportation Research Record: Journal of the Transportation Research Board*, 2004. 1867: 172–182.
7. Liu, H. X., X. Wu, and P. G. Michalopoulos. Improving Queue Size Estimation for Minnesota's Stratified Zone Metering Strategy. *Transportation Research Record: Journal of the Transportation Research Board*, 2007. 2012: 38–46.
8. Vigos, G., M. Papageorgiou, and Y. Wang. Real-Time Estimation of Vehicle-Count within Signalized Links. *Transportation Research Part C: Emerging Technologies*, Vol. 16, No. 1, 2008, pp. 18–35.
9. Wu, J., X. Jin, and A. J. Horawitz. Methodologies for Estimating Vehicle Queue Length at Metered On-Ramps. *Transportation Research Record: Journal of the Transportation Research Board*, 2008. 2047: 75–82.
10. Sanchez, R. O., R. Horawitz, and P. Varaiya. Analysis of Queue Estimation Methods Using Wireless Magnetic Sensors. *Transportation Research Record: Journal of the Transportation Research Board*, 2011. 2229: 34–45.
11. Lee, I., R. Jiang, and E. Chung. Traffic Queue Estimation for Metered Motorway On-Ramps through Use of Loop Detector Time Occupancies. *Transportation Research Record: Journal of the Transportation Research Board*, 2013. 2396: 45–53.
12. Yang, G., Z. Tian, H. Xu, and D. Wang. Queue Length Estimation at Metered Freeway-to-Freeway Connectors. *Journal of Advanced Transportation*, Vol. 50, No. 8, 2016, pp. 1912–1924.
13. Yang, G., Z. Tian, H. Xu, Z. Wang, and D. Wang. Impacts of Traffic Flow Arrival Pattern on the Necessary Queue Storage

Space at Metered On-Ramps. *Transportmetrica A: Transport Science*, 2017, pp. 1–19. https://doi.org/10.1080/23249935.2017.1387875.

14. Akcelik, R., and N. M. Rouphail. Overflow Queues and Delays with Random and Platoon Arrival at Signalized Intersections. *Journal of Advanced Transportation*, Vol. 28, No. 3, 1994, pp. 227–251.
15. Lin, F., P. Tseng, and C. Su. Variations in Queue Discharge Patterns and Their Implications in Analysis of Signalized Intersections. *Transportation Research Record: Journal of the Transportation Research Board*, 2004. 1883: 192–197.
16. Geroliminis, N., and A. Skabardonis. Prediction of Arrival Profiles and Queue Lengths along Signalized Arterials by using a Markov Decision Process. *Transportation Research Record: Journal of the Transportation Research Board*, 2005. 1934: 116–124.
17. Bonneson, J., M. Pratt, and M. Vandehey. Predicting Arrival Flow Profiles and Platoon Dispersion for Urban Street Segments. *Transportation Research Record: Journal of the Transportation Research Board*, 2010. 2173: 28–35.
18. Newell, G. F. *Applications of Queuing Theory*, 2nd ed. Chapman and Hall, New York, 1982.
19. Tian, Z. *Development and Evaluation of Operational Strategies for Providing an Integrated Diamond Interchange Ramp-Metering Control System*. PhD dissertation, Texas A&M University, College Station, Tex., 2004.

The Standing Committee on Highway Capacity and Quality of Service (AHB40) peer-reviewed this paper (18-05940).

Article

TRR
JOURNAL OF THE TRANSPORTATION RESEARCH BOARD

Transportation Research Record
2018, Vol. 2672(15) 93–104

DOI: 10.1177/0361198118798723
journals.sagepub.com/home/trr
SAGE

Validation and Calibration of Freeway Reliability Methodology in the Highway Capacity Manual: Method and Case Studies

Nabaruna Karmakar[1], Seyedbehzad Aghdashi[2], Nagui M. Rouphail[3], and Billy M. Williams[3]

Abstract

Traffic congestion costs drivers an average of $1,200 a year in wasted fuel and time, with most travelers becoming less tolerant of unexpected delays. Substantial efforts have been made to account for the impact of non-recurring sources of congestion on travel time reliability. The 6th edition of the *Highway Capacity Manual* (HCM) provides a structured guidance on a step-by-step analysis to estimate reliability performance measures on freeway facilities. However, practical implementation of these methods poses its own challenges. Performing these analyses requires assimilation of data scattered in different platforms, and this assimilation is complicated further by the fact that data and data platforms differ from state to state. This paper focuses on practical calibration and validation methods of the core and reliability analyses described in the HCM. The main objective is to provide HCM users with guidance on collecting data for freeway reliability analysis as well as validating the reliability performance measures predictions of the HCM methodology. A real-world case study on three routes on Interstate 40 in the Raleigh-Durham area in North Carolina is used to describe the steps required for conducting this analysis. The travel time index (TTI) distribution, reported by the HCM models, was found to match those from probe-based travel time data closely up to the 80th percentile values. However, because of a mismatch between the actual and HCM estimated incident allocation patterns both spatially and temporally, and the fact that traffic demands in the HCM methods are by default insensitive to the occurrence of major incidents, the HCM approach tended to generate larger travel time values in the upper regions of the travel time distribution.

Increasing traffic congestion and driver frustration on urban freeways have been the motivation behind several studies related to travel time reliability. Most travelers are less tolerant of unexpected delays because they have larger consequences than everyday congestion (*1*). Traffic jams cost drivers in the U.S. an average of $1,200 a year in wasted fuel and time, according to a study by INRIX (*2*). Travel time reliability is an important performance measure used to describe travel characteristics on freeways and urban streets. It is used to quantify variability in travel time, which can be a result of both recurring and non-recurring congestion such as inclement weather or incidents. Recent federal rulemakings published by the Federal Highway Administration (FHWA) have proposed that reliability performance measures be used as a basis for federal funding prioritization of road projects (*3*).

Substantial efforts have been made to account for the impact of non-recurring sources of congestion on travel time variation. The *Highway Capacity Manual* (HCM), one of the most widely used references used by analysts for evaluating capacity and level of service on freeways, also provides methods to account for the impact of non-recurring sources of congestion (*4*). Chapters 11 and 17 of the latest release of the manual contain a methodology for travel time reliability analysis for freeways and arterial streets. This entails expanding the analysis window from a few hours on a single day to a whole year. Non-

[1]Edward P. Fitts Department of Industrial and Systems Engineering, NC State University, Raleigh, NC
[2]Institute for Transportation Research and Education, NC State University, NC
[3]Department of Civil, Construction, and Environmental Engineering, NC State University, Raleigh, NC

Corresponding Author:
Address correspondence to Nabaruna Karmakar: nkarmak@ncsu.edu

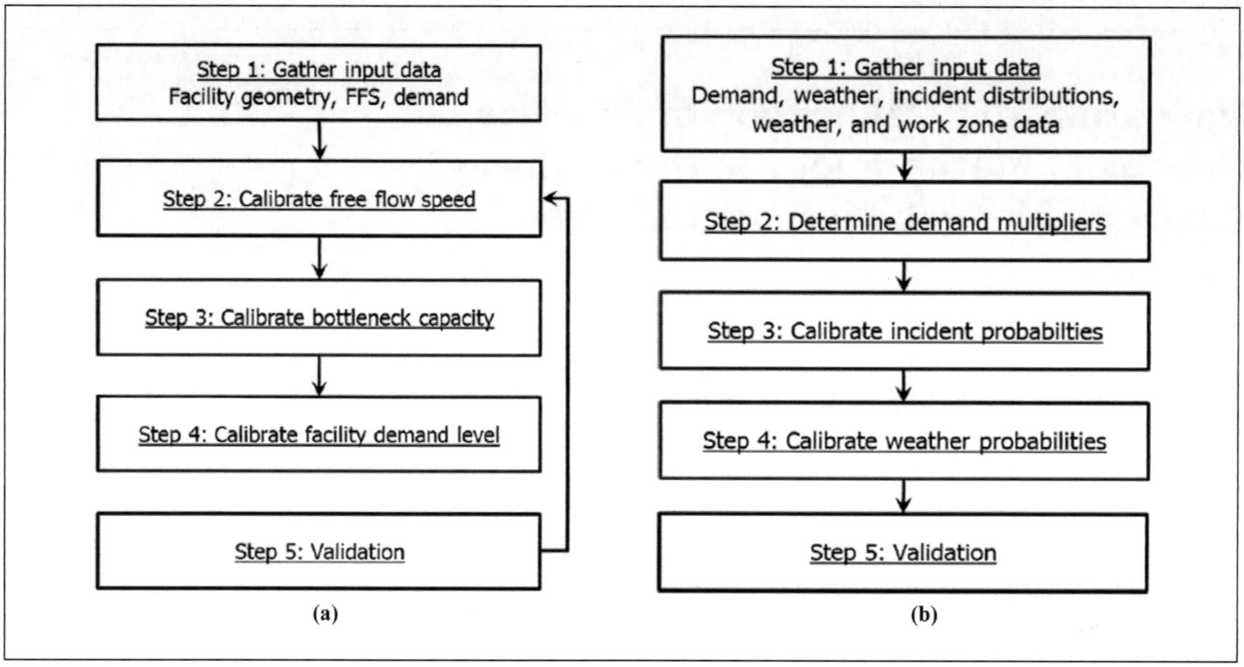

Figure 1. Calibration process for freeway core and reliability analysis in HCM (*4*). (*a*) HCM exhibit 25-26 core analysis calibration procedure; (*b*) HCM exhibit 25-32 reliability calibration procedure.

recurring sources of congestion such as inclement weather, incidents, and work zones are accounted for within this wider window. By default, the HCM freeway reliability methodology generates 240 scenarios (roughly equivalent to the number of work weekdays in a year) to simulate variations in demand, weather, and incident events throughout the year (*5*).

A key prerequisite to carrying out a reliability analysis is a single day (or seed day) calibration of the facility being analyzed. The seed-day scenario is then augmented to include recurring (demand) and non-recurring sources of congestion, thus generating new and different sets of scenarios. As the goal is to represent real-world operating conditions on the facility, all sources of congestion must also be calibrated using field data. This can be done through matching the likelihoods of events like incidents, severe weather, work zones, and so forth, in the model to actual observations and their impact on travel time.

The 6th edition of HCM provides structured guidance on a step-by-step calibration of the core (single day) as well as reliability analysis, as displayed in Figure 1. However, practical implementation of these methods requires collecting data scattered in different platforms and available in inconsistent formats. The most widely used data source to calibrate the core analysis are probe-based speed data (e.g., from INRIX, HERE). The reliability calibration requires weather information daily incident frequencies, which must be extracted and filtered for errors, omissions, and outliers, all of which can affect the incident frequency estimations.

This paper focuses on the calibration and validation of both the core (single day) and reliability analyses in HCM (*4*). The main objective is to provide HCM users with guidance on collecting data for freeway reliability analysis and validating the reliability performance measures predictions of HCM models. An extensive review of research pertaining to travel time reliability modeling is provided, and the process for collecting and pre-processing the inputs required for such analysis is explained. The output of reliability analysis in the HCM method is then validated using a case study on Interstate 40 near Raleigh-Durham, NC.

Literature Review

Travel time reliability includes the notion of predictability, that is, the probability that the travel time on a facility is within acceptable limits for the traveler, given that travel times are affected by interaction of demand fluctuations, traffic control devices, traffic incidents, inclement weather, work zones, and physical capacity. A variety of performance measures have been proposed to quantify reliability, which include 80th percentile travel time, travel time index (TTI), planning time index (PTI), buffer time index, and many others. The use of an indexed term like TTI or PTI is generally preferred as it normalizes the travel time by the free-flow travel time on the facility, which enables easier comparison across facilities. Among these measures, a National Cooperative Highway Research Program effort identified the *Buffer Time Index* as the most useful reliability performance measure, including a successful field test by four U.S. transportation agencies (*6*). Franklin modeled reliability as *expected lateness* using a schedule-based approach (*7*). The study examined travel time data for mean lateness, an important component of scheduling, to produce a complete estimate of user benefit.

Rakha et al. investigated the relationships between segment and trip travel time reliability metrics, and explored and validated options for characterizing trip reliability (*8*). Through another large national research effort conducted under the SHRP-2 L03 project, many reliability performance measures and macroscopic methods for estimating key reliability measures from average travel times were proposed (*9*). Studies have investigated the impact of stochastic bottleneck capacity on reliability (*10*) and even extended to estimating travel time distribution based on demand and capacity (*11, 12*).

Traffic demand variations, inclement weather conditions, and incidents are recognized as principal factors affecting the variability of travel times on the freeways (*13, 14*). Presence of construction zones have been studied as one of the sources of non-recurring congestion. Al Kaisy and Hall developed guidelines for estimating freeway capacity at long-term reconstruction zones (*15*), and Krammes and Lopez (*16*) did the same for short-term freeway work zones. Chituri et al. (*17*) and Schroeder and Rouphail (*18*) extended those analyses to include delay and user costs for freeway work zones.

Similarly, several researchers have studied the impact of weather on non-recurring congestion. A change in weather conditions can affect the capacity and the free-flow speed of the freeway, resulting in changes in travel time (*13, 19*). The occurrence of different weather conditions is difficult to predict, but their expectancy could be estimated based on historical averages (*20*). Rakha et al. proposed specific adjustments to traffic flow parameters based on specific weather events (*21*). Similar work on the effects of rainfall and other environmental factors on traffic flow characteristics on urban freeways have been reported by Shi et al. (*22*), whereas Dehman (*23*) focused on the effect of inclement weather on both capacity and queue discharge headways at recurring freeway bottlenecks.

The occurrence of incidents is stochastic in nature and can result in a reduction in the capacity and/or speeds (*4, 13, 24*). Moreover, the location, severity, and duration of an incident can altogether influence its impact on the facility travel time. Skabardonis et al. showed that incident occurrence on a freeway facility follows the Poisson distribution (*25*). However, more detailed research has suggested that incident occurrence follows a binomial distribution and that incident counts are much more dispersed, with variance greater than the mean (*14*). Research involving distribution of duration of incidents on freeways assumes that it closely follows a Lognormal or a Gamma distribution (*14*).

From the literature, it is evident that weather, work zones, and incidents are all likely to have significant impact on freeway capacity, free-flow speeds, or in many cases both. The latest version of HCM comes with a revised method to generate reliability scenarios and calculate performance metrics, primarily TTI distributions. However, validations conducted as part of the SHRP 2 L08 project revealed that the observed capacity values in the field were usually lower than the proposed values in the manual (*26–30*).

During the core facility calibration, if an unwanted bottleneck appeared in the model, the guidance recommends increasing the capacity of a segment until it disappears. Beyond this, the HCM provides no insight on how to adjust capacity and match predicted speeds to field-observed speeds. Therefore in many cases, the capacity calibration process becomes a trial-and-error process until a satisfactory level of accuracy is achieved. A new optimization-based approach for bottleneck calibration on freeway facilities, in the context of the HCM analysis, uses a genetic algorithm to adjust segment demand and capacity to match HCM-modeled and observed speeds predicted at a high level of confidence (*31*).

Methodology

This section of the paper discusses the implementation of both the core and reliability analyses in the HCM, along with the application of the proposed methodologies on three selected study routes.

Site Selection

The temporal and spatial boundaries of the study routes were carefully considered in order to capture daily recurring congestion patterns as well as all seasonal travel trends. Two freeway sections on I-40 in the Raleigh-Durham area in North Carolina, selected for reliability analysis, are shown in Figure 2.

Route A, an urban interstate route, is 14.72 miles long along I-40 Eastbound from US-1 to NC-147, contains 23 traffic message channel (TMC) segments, and experiences a morning peak demand. It is primarily a commuter route that passes through the Research Triangle Park, a major employment center in the area and has a speed limit of 65 mph along the entire path. Route B is the westbound direction of travel on this route, with a p.m.

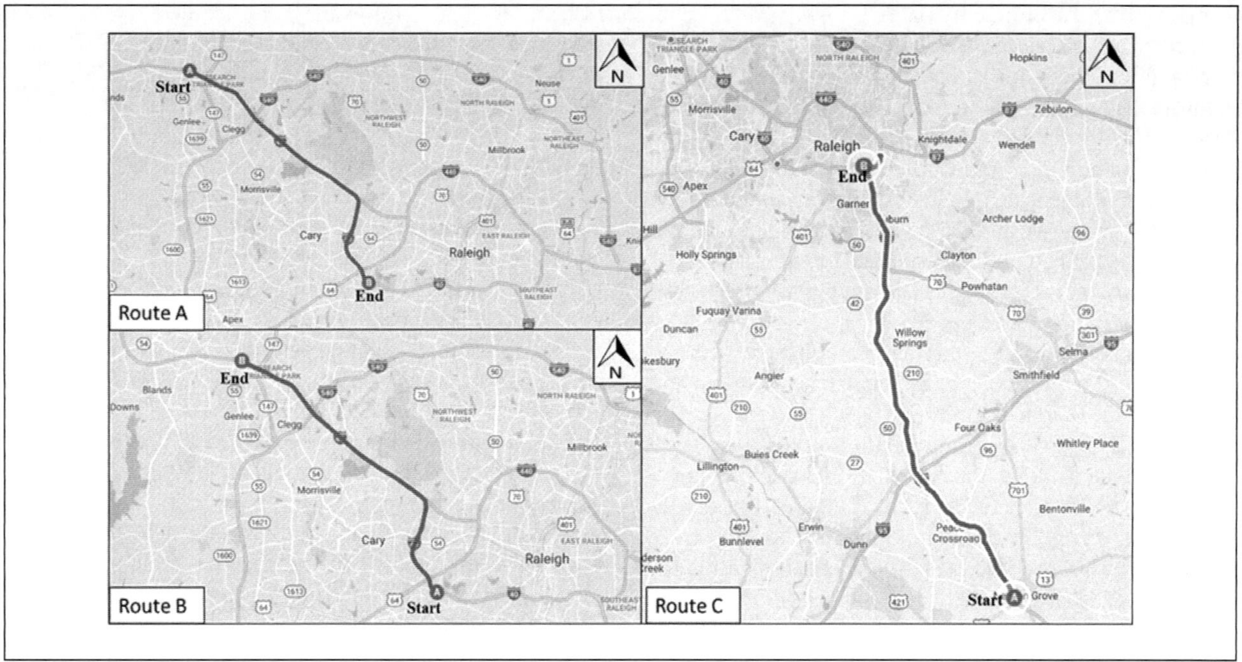

Figure 2. Directional (A→B) freeway facilities analyzed.

peak travel. Route C, an exurban/ rural interstate, is 39.84 miles long on I-40 WB from NC-50/55 to I-440 and contains 17 TMC links. The path crosses I-95 on the way to Raleigh, NC from southeastern North Carolina. The speed limit is 70 mph from the origin until approximately 30 miles into the path, where the speed limit drops to 65 mph for the remainder of the path. This route experiences an a.m. peak commuting period. A study period of 24-hour and a reliability reporting period included all weekdays for the year 2015.

Core Methodology Calibration and Modeling

This section discusses the steps needed to calibrate the core facility (see also Figure 1*a*).

Step 1: Facility Geometry. The first step involves defining the segments' geometry consistent with the requirements of performing the analysis method described in the HCM. In this study, routes A, B, and C consist of 37, 45 and 37 HCM segments, respectively, per the HCM guidelines on segmenting freeway facilities. Other inputs such as segment length, number of lanes, shoulder clearances, and so forth, were extracted directly from Google Maps.

Step 2: Calibrate Free-Flow Speed. Travel times at very low traffic demand conditions represent free-flow travel times, as a direct function of free-flow speed. The estimated free-flow travel times at low traffic demand flows were compared with the probe-based sensor data and necessary adjustments were made to calibrate the free-flow speed in the models.

Step 3: Calibrate Bottleneck Capacity. An automated approach for the HCM bottleneck capacity calibration process was used in this step (*31*). It involves mathematical optimization, using a genetic algorithm metaheuristic to estimate capacity adjustment factors and identify facility. It optimizes the capacity adjustments such that they result in predicted performance measures consistent with the collected empirical speed data and significantly helps reduce the time and effort needed for analysis.

Step 4: Calibrate Facility Demand Level. Demand flow rates for the mainline, on-ramp entry, and off-ramp exit were estimated based on annual average daily traffic (AADT) and hourly demand profiles, and subsequently aggregated into 15-minute intervals. Collectively, this forms the basis of the input for the model.

Step 5: Validation. The primary data source for calibration was INRIX (extracted from RITIS.org), which provides probe-based travel time and speed data (*32*). Several days' worth of data were gathered, and outlier days, such as days with unusual congestion patterns or no congestion, were excluded from the dataset. INRIX uses internal and external TMC codes as the smallest spatial unit

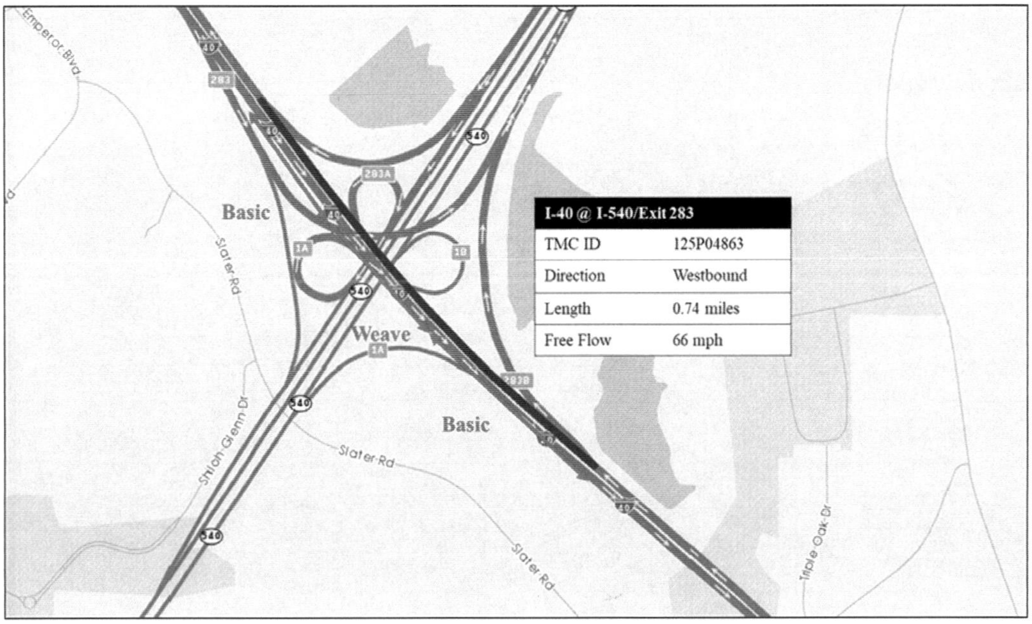

Figure 3. An INRIX TMC segment on WB I-40 (in purple) containing 3 HCM segments (in red) at the interchange with I-540.

for reporting freeways performance measures. This method reports traffic data between each break in access on any road (such as from one on-ramp to next off-ramp). It was observed that a single TMC segment usually covers two or more HCM segments, as per the example shown in Figure 3.

To perform validation via probe-based data sources, the actual facility as configured was converted into HCM segments. A length-based average of HCM segments' performance measures is used to match TMC segments. Thus, if a TMC segment contains n HCM segments of length L_i each, then the aggregated travel time performance measure is computed as:

$$TT_{\text{TMC}} = \frac{\sum_{i=1}^{n} L_i \times TT_{\text{HCM},i}}{\sum_{i=1}^{n} L_i} \quad (1)$$

Where TT_{TMC} is travel time on TMC segment and TT_{HCM} is travel time on HCM segment.

Equation 1 was used to match the performance of TMC and HCM segments. This process provided the basis for comparing the HCM reported performance measures with TMC segment observations.

Reliability Analysis Calibration and Modeling

After the core calibration of the three selected facilities, calibration for the reliability scenarios was performed. As stated in HCM Chapter 11, characterization of the demand fluctuation across days of the week and months of the year, along with calibration of non-recurring congestion sources, that is, incidents, weather, work zones, and so forth (*4*). The following steps are based on Figure 1*b*.

Step 1: Gather Input Data. The types of data required for calibration of reliability analysis include demand, weather, incident distributions, and work zone data, as shown in Figure 1*b*. Chapter 11 of the HCM also provides information for data requirements for agencies with low access to good quality data (*4*). The best source of demand fluctuation data is permanent traffic recorders located alongside the facility. Demand multipliers are a direct input into the HCM model and represent the ratio of the traffic demand in each analysis period to the AADT. Those are then used to generate the input demand level for each scenario. Calibrating weather events requires historical weather data, which were gathered from the Weather Underground website. For modeling the effect of incidents, detailed incident logs are needed, which were collected from incident management websites managed by North Carolina Department of Transportation (NCDOT). It should be noted that the study sites in this paper did not have any on-going construction zones during the year 2015.

Step 2: Determine Demand Multipliers. Demand level should be studied for the facility where the reliability analysis is

Table 1. Demand Multipliers on Interstate 40

Demand multiplier		Day of week				
		Monday	Tuesday	Wednesday	Thursday	Friday
Month of year	January	0.99662	1.02778	1.04039	1.0526	1.08161
	February	0.93925	1.01073	1.03921	1.09203	1.14007
	March	1.04331	1.06934	1.06352	1.11092	1.17112
	April	1.07358	1.08746	1.09824	1.16197	1.215
	May	1.07633	1.10618	1.11396	1.15772	1.21043
	June	1.07804	1.08585	1.06747	1.13872	1.18033
	July	1.08258	1.07099	1.10251	1.14728	1.18498
	August	1.04605	1.05215	1.06037	1.09324	1.1649
	September	1.01602	1.02405	1.02363	1.07478	1.15295
	October	1.04898	1.04572	1.06699	1.10704	1.16095
	November	0.97404	0.99995	1.04121	1.08154	1.07035
	December	0.97479	0.95648	0.98702	0.91611	1.0077

performed (*13*). Categorization of demand is done by assigning similar demand patterns to specific days with the same demand level in the reliability scenario generator. Demand patterns are defined along two dimensions, accounting for monthly and weekday variability. Monthly variability accounts for seasonal trends in travel, whereas the weekday dimension shows the effect of day-to-day variation in demand. These demand multipliers give the ratio of demand for a day–month combination to the AADT, and are used to generate demand values in the reliability scenario generation. As the subject facilities are quite close to each other, the permanent counts station data were aggregated for I-40 to construct Table 1 and the resulting demand multipliers were used for all three sites.

Step 3: Calibrate Incident Probabilities. The HCM method requires, as input, incident frequencies by month of year, and incident duration statistics categorized by severity. Characterizing incidents requires cleaning and processing of annual incident log datasets, typically maintained by state DOTs or other roadway agencies. Such databases usually contain incident reports, and include attributes such as start and end time, road name, direction, mile marker, severity, number of closed lanes, and many more. After these data are collected for the specific freeway being analyzed, they should be filtered spatially using mile markers and temporally using the reported start times and confined to the scope of study. The duration of an incident is calculated as the difference between its reported start and end time. As incident logs are usually manually reported and, therefore, susceptible to human error, outlier removal is a necessary step.

For the selected routes, incident records were collected from NCDOT Travel Information Management Systems (TIMS) website (*33*). All incident observations on I-40 for weekdays in calendar year 2015 were filtered by mile markers and direction for each route. Then, outlier observations were removed using an inter-quartile range check on incident duration grouped by severity type (shoulder closure, one lane closure, etc.). The records were then processed to calculate daily incident frequencies for each month of the year, and incident duration statistics including the mean, standard deviation, min and max were computed and categorized by severity. These were fed into the HCM model, which randomly simulates incident occurrence and durations in an analysis period using Poisson and Log Normal distributions, respectively, fitted to the input parameters. The start time and location of these incidents are assigned using the daily vehicle miles traveled (VMT) profile on each segment. Tables 2 and 3 show the daily incident frequencies, incident duration statistics, and distribution between different closure types on the three study facilities, which serve as an input to the HCM model.

Step 4: Calibrate Weather Probabilities. The HCM methodology also requires the specification of the probability of up to 11 different weather events by calendar month as input. Calibrating weather events requires historical weather data, which are used to estimate the probability, average duration, and standard deviation of duration of different weather events. In this paper, the research team used regional default weather event probabilities and default impacts. The HCM computational engine has a database of 10 years' average of weather events probabilities for about a hundred largest metropolitan areas in the U.S., via the Weather Underground service (*34*). This study used the Raleigh-Durham Airport location to extract weather probabilities for the case studies.

Step 5: Validation. The results of the model were validated against travel times collected for the corresponding TMC segments on each route. As the HCM method uses 15-minute intervals as the basic temporal unit of analysis, 15-minute average reported TMC travel times were gathered from INRIX. These travel times were converted to a route TTI using a simultaneous path travel time estimation method (*35*). The TTI distribution of the HCM method was then validated against that counted from INRIX speed data.

Table 2. Incident Daily Frequencies for Route A, Route B, and Route C in 2015

Month	Number of incidents	Number of days	Incident frequencies
Route A			
January	29	22	1.3181818
February	38	20	1.9
March	26	22	1.1818182
April	28	22	1.2727273
May	23	21	1.0952381
June	25	22	1.1363636
July	22	23	0.9565217
August	38	21	1.8095238
September	27	22	1.2272727
October	30	22	1.3636364
November	37	21	1.7619048
December	26	23	1.1304348
Route B			
January	28	22	1.272727
February	40	20	2
March	31	22	1.409091
April	32	22	1.454545
May	22	21	1.047619
June	26	22	1.181818
July	24	23	1.043478
August	37	21	1.761905
September	30	22	1.363636
October	42	22	1.909091
November	34	21	1.619048
December	32	23	1.391304
Route C			
January	11	22	0.5
February	20	20	1
March	7	22	0.31818182
April	7	22	0.31818182
May	9	21	0.42857143
June	12	22	0.54545455
July	1	23	0.04347826
August	3	21	0.14285714
September	8	22	0.36363636
October	10	22	0.45454545
November	11	21	0.52380952
December	7	23	0.30434783

Results

Key Reliability Performance Measures

Figure 4 shows the various distribution statistics, alongside the cumulative distribution functions (CDFs) of the emerging TTI distribution estimated by the HCM reliability method and the empirical distribution of the INRIX travel times for the corresponding TMC set, for each of the 15-minute analysis periods over all weekdays in the year 2015. The average, 50th percentile and 80th percentile TTIs computed using the HCM method for all three routes were all within 15% of those reported by INRIX. FHWA has proposed new reliability performance measures, such as level of travel time reliability (LOTTR), as part of the Moving Ahead for Progress in the 21st Century Act (MAP-21) (*36*). These metrics were calculated from the HCM method outputs and were found to be within 15% of INRIX observations.

Table 3. Incident Duration Statistics for Route A, Route B, and Route C

Route A

Incident type	Count	Average duration (min)	Std dev duration (min)	Max duration (min)	Min duration (min)	Percentage distribution (%)
1 Lane closed	14	25.1	17.4	52.4	8.2	22.2
2 Lane closed	3	60.4	58.3	123.5	8.7	4.8
3 Lane closed	0	0.0	0.0	0.0	0.0	0.0
Shoulder closed	46	63.1	32.8	136.9	11.7	73.0

Route B

Incident type	Count	Average duration (min)	Std dev duration (min)	Max duration (min)	Min duration (min)	Percentage distribution (%)
1 Lane closed	22	27.3	20.3	75.5	6.1	18.8
2 Lane closed	6	38.6	30.9	92.2	11.1	5.1
3 Lane closed	0	0.0	0.0	0.0	0.0	0.0
Shoulder closed	89	56.9	34.3	151.2	5.2	76.1

Route C

Incident type	Count	Average duration (min)	Std dev duration (min)	Max duration (min)	Min duration (min)	Percentage distribution (%)
1 Lane closed	8	171.5	182.6	540.0	28.5	17.0
2 Lane closed	0	0.0	0.0	0.0	0.0	0.0
Shoulder closed	39	49.9	24.0	108.7	6.1	83.0

However, for Route A, the evening peak LOTTR values were significantly higher than observations from INRIX data. Similarly, the higher region of the distribution at or above the 90th percentile TTI for Route A showed significant differences between the estimated and observed CDFs. For routes B and C, the overall CDFs and performance statistics were very consistent between the HCM method and empirical data. A two-sample Kolmogorov–Smirnoff (KS) test conducted on the three sets of CDFs, shown in Figure 4, showed that the distributions predicted by the HCM method were not significantly different from those observed from INRIX data at the 5% significance level. However, the two-sample KS test is not robust to difference in the tail end of distributions, which could be the reason it is unable to find differences between the CDFs in Figure 4*a*. The next section looks more closely at the incident scenario generation technique to investigate possible contributions to the large discrepancy in the upper regions of the TTI distribution for Route A.

Discussion

To further investigate the higher estimated TTIs at and above the 90th percentile and the evening peak hour on Route A, incidents generated by the HCM method over time of day were compared with the TIMS log data. Figure 5 shows the total number of incident occurrence based on the HCM method, and in the incident logs, by hour of day. As shown in Figure 5*a*, the HCM model does not adequately capture the temporal incident trends found in the actual data (Figure 5*b*). This is primarily the result of the assumption in the HCM to use VMT distribution to allocate incident start time and location in the reliability scenarios. It is curious, however, that this pattern generated higher values of the 95th percentile TTI in the HCM model, as was shown in Figure 4*a*. One possible explanation is that the accumulation of incidents starting in the early afternoon, based on the HCM model allocation, may have initiated congestion earlier, which was exacerbated as traffic demands kept increasing during the p.m. peak period.

Another explanation of the discrepancy may be attributed to the discretization of all incident durations in the HCM model to the nearest 15 minutes. Hence, all actual incidents that clear in fewer than 15 minutes are not considered in the HCM approach. In effect, the method will tend to consistently model incidents with duration higher than empirical observations indicate. This pattern can be detected in Figure 6, which shows the incident duration distributions in both HCM and in TIMS. Finally, the HCM model does not consider any changes in traffic demand (i.e., as a result of diversions) in response to major incidents, which may also result in inflating their travel time effect when generating the TTI distribution.

TTI Metrics	ROUTE A HCM Method	ROUTE A INRIX
Average	1.31	1.13
Min	1.02	0.95
Max	22.79	6.79
Std Dev	0.77	0.34
50th Percentile	1.06	1.03
80th Percentile	1.12	1.06
95th Percentile	3.04	1.29
LOTTR (6 am to 10 am)	1.02	1.10
LOTTR (10 am to 4 pm)	1.03	1.02
LOTTR (4 pm to 8 pm)	1.65	1.19

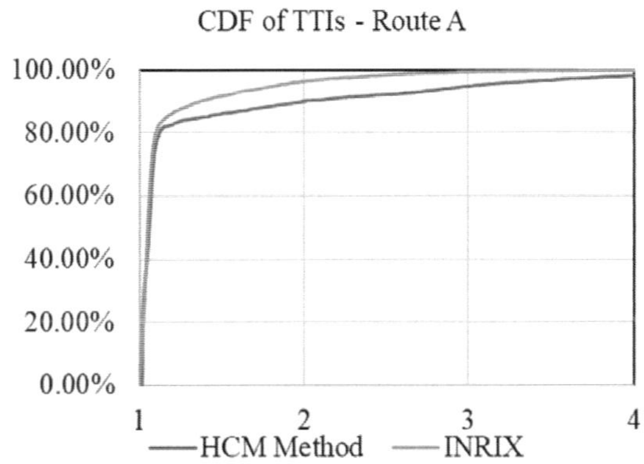

(a)

TTI Metrics	ROUTE B HCM Method	ROUTE B INRIX
Average	1.11	1.09
Min	1.02	0.95
Max	8.39	5.72
Std Dev	0.24	0.22
50th Percentile	1.06	1.03
80th Percentile	1.09	1.06
95th Percentile	1.46	1.29
LOTTR (6 am to 10 am)	1.05	1.01
LOTTR (10 am to 4 pm)	1.02	1.01
LOTTR (4 pm to 8 pm)	1.54	1.39

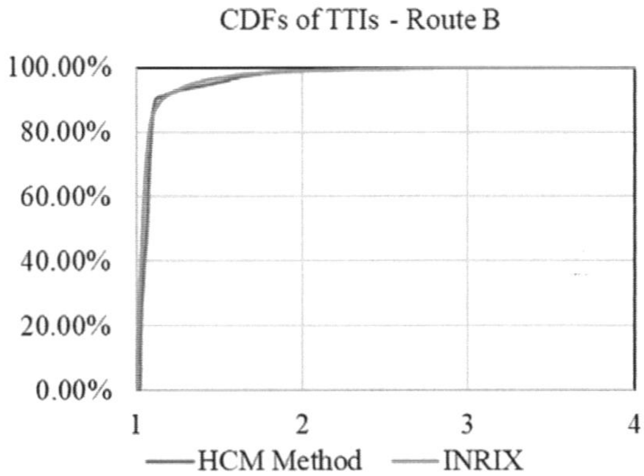

(b)

TTI Metrics	ROUTE C HCM Method	ROUTE C INRIX
Average	1.04	1.02
Min	1.02	0.95
Max	13.63	2.41
Std Dev	0.18	0.09
50th Percentile	1.02	1.00
80th Percentile	1.02	1.02
95th Percentile	1.20	1.12
LOTTR (6 am to 10 am)	1.09	1.11
LOTTR (10 am to 4 pm)	1.00	1.02
LOTTR (4 pm to 8 pm)	1.00	1.02

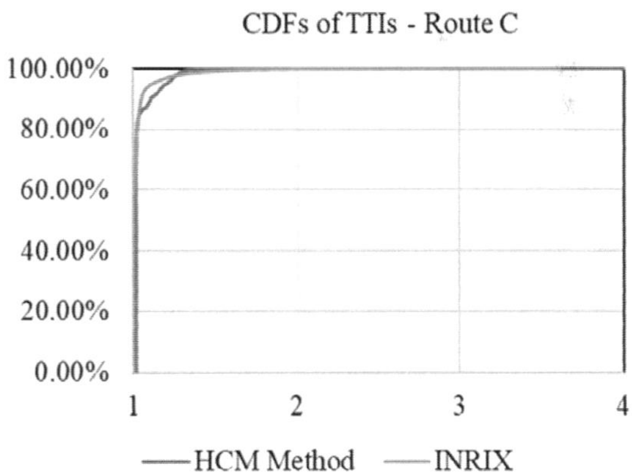

(c)

Figure 4. TTI reliability statistics and CDF from HCM and INRIX by route. (*a*) Route A; (*b*) Rote B; (*c*) Route C.

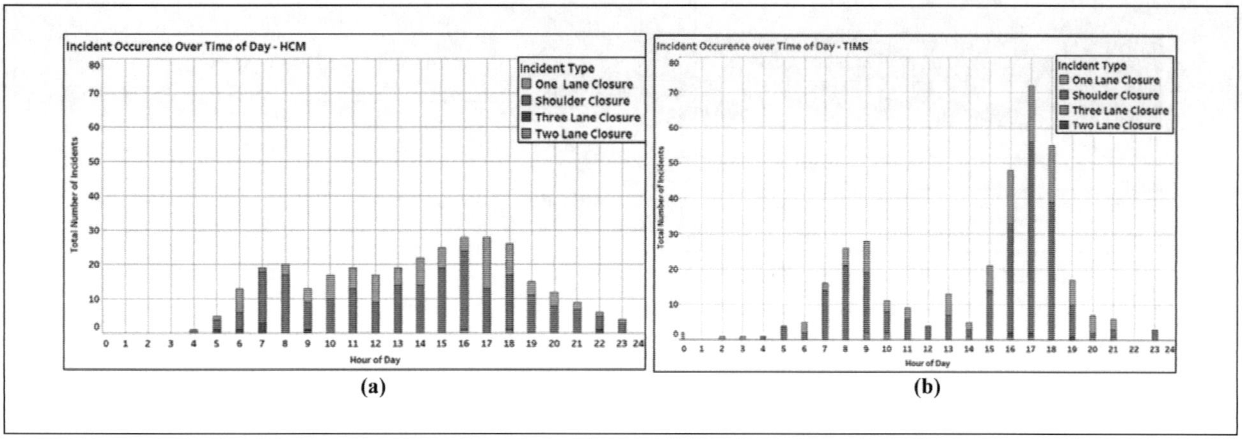

Figure 5. Incidents by time of day on Route A in (*a*) HCM method and (*b*) TIMS logs.

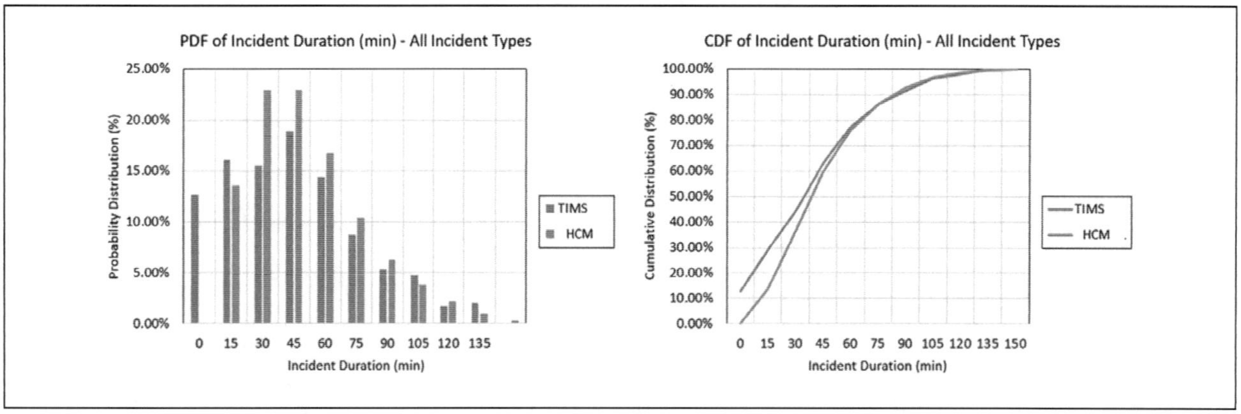

Figure 6. Distributions of incident duration: TIMS vs. HCM.

Conclusions and Future Work

This paper presented a case study implementation of the freeway reliability analysis procedure in the 6th edition of the HCM. The major challenges involved in such implementation were identified and explained. The data needed for performing this analysis were explained and the results of the reliability analysis of the study sites were validated. The TTI distribution up to the 80th percentile value was found to match closely those observed from probe-based travel times. However, the upper tail of the distribution from the HCM method yielded very high travel times. This can be explained by the fact that the HCM method models incidents with higher average duration than is seen in the empirical data, which exacerbates the poor travel conditions along the facility, especially when there are multiple incidents.

The incidents modeled in the HCM method follow the distribution of the VMT in a 24-hour period. However, for urban road segments, during peak periods, incidents can have a cascading effect as traffic demands increase when approaching the peak periods. The effect of one incident in a segment could lead to later secondary incidents on upstream segments. Further research is needed to investigate the probability of secondary incidents occurring on a facility as a direct effect of a downstream incident. With the use of mobile devices, drivers can choose to alternate routes more easily to avoid congestion caused by incidents. The HCM method does not take driver route choice or diversion into consideration when modeling the effect of incidents, thus probably inflating their effect. Lastly, the reliability scenario generation described in the HCM assumes that the effect of weather events and incidents are mutually independent. However, the input parameters for incident estimation in the model are not free from such an assumption. Future work is needed to incorporate the interaction of weather events and incidents on freeways for a more realistic modeling of reliability.

Acknowledgments

The research presented in this paper was funded by the NCDOT and FHWA through NCDOT project 2016-32 titled "SHRP2 Reliability Data and Analysis Tools: Implementation Assistance Program Pilot Study." The research team is grateful for the support it has received from NCDOT and FHWA.

Author Contributions

The authors confirm contribution to the paper as follows: study conception and design: Nabaruna Karmakar, Seyedbehzad Aghdashi, Nagui M. Rouphail, Billy M. Williams; data collection: Nabaruna Karmakar, Seyedbehzad Aghdashi; analysis and interpretation of results: Nabaruna Karmakar, Seyedbehzad Aghdashi, Nagui M. Rouphail, Billy M. Williams; draft manuscript preparation: Nabaruna Karmakar, Seyedbehzad Aghdashi, Nagui M. Rouphail, Billy M. Williams. All authors reviewed the results and approved the final version of the manuscript.

References

1. FHWA Office of Operations. *Travel Time Reliability: Making It There on Time, All the Time*. https://ops.fhwa.dot.gov/publications/tt_reliability/TTR_Report.htm. Accessed July 26, 2017.
2. Joseph, W. *Traffic Jams Cost U.S. Drivers $1,200 a Year: Study*. Reuters. 2017. https://www.usnews.com/news/us/articles/2017-02-20/traffic-jams-cost-us-drivers-1-200-a-year-study. Accessed July 27, 2017.
3. Federal Register 2016-08014. *National Performance Management Measures; Assessing Performance of the National Highway System, Freight Movement on the Interstate System, and Congestion Mitigation and Air Quality Improvement Program*. FHWA, U.S. Department of Transportation, Washington, D.C., 2016.
4. *Highway Capacity Manual: A Guide for Multimodal Mobility Analysis*, 6th ed. Transportation Research Board, Washington, D.C., 2016.
5. Aghdashi, S., A. Hajbabaie, B. J. Schroeder, J. L. Trask, and N. M. Rouphail. Generating Scenarios of Freeway Reliability Analysis: Hybrid Approach. *Transportation Research Record: Journal of the Transportation Research Board*, 2015. 2483: 148–159.
6. Tarnoff, P., S. Young, J. Crunkleton, and N. Nezamuddin, and University of Maryland. *NCHRP 20-7, Task 202: Guide to Benchmarking Operations Performance Measures*. Transportation Research Board of the National Academies, Washington, D.C., 2008.
7. Franklin, J. P. Modeling Reliability as Expected Lateness: A Schedule-Based Approach for User Benefit Analysis, Royal Institute of Technology, Stockholm, Sweden. *Proc., European Transport Conference*, Noordwijkerhout, Netherlands, 2009.
8. Rakha, H., I. El-Shawarby, and M. Arafeh. Trip Travel-Time Reliability: Issues and Proposed Solutions. *Journal of Intelligent Transportation Systems*, Vol. 14, No. 4, 2010, pp. 232–250. http://10.1080/15472450.2010.517477.
9. Cambridge Systematics, Texas A&M University, University of Washington, Dowling Associates, Street Smarts, Herb Levinson and Hesham Rakha. *Analytical Procedures for Determining the Impacts of Reliability Mitigation Strategies*. SHRP 2 Report S2-L03-RR-1. 2013. https://doi.org/10.17226/22806.
10. Brilon, W., J. Geistefeldt, and H. Zurlinden. Implementing the Concept of Reliability for Highway Capacity Analysis. *Transportation Research Record: Journal of the Transportation Research Board*, 2007. 2027: 1–8.
11. Jia, A., B. Williams, and N. Rouphail. Identification and Calibration of Site-Specific Stochastic Freeway Breakdown and Queue Discharge. *Transportation Research Record: Journal of the Transportation Research Board*, 2010. 2188: 148–155.
12. Zhou, X., N. Rouphail, and M. Li. Analytical Models for Quantifying Travel Time Variability Based on Stochastic Capacity and Demand Distributions, Presented at 90th Annual Meeting of the Transportation Research Board, Washington, D.C., 2011.
13. Kittleson and Associates. *SHRP 2 Report S2-L08-RW-1: Incorporating Travel Time Reliability Analysis into Highway Capacity Manual*. Transportation Research Board of the National Academies, Washington, D.C., 2013.
14. Hajbabaie, A., S. Aghdashi, and N. Rouphail. Enhanced Decision-Making Framework Using Reliability Concepts for Freeway Facilities. *Journal of Transportation Engineering*, Vol. 142, No. 4, 2016, p. 04016008. https://doi.org/10.1061/(asce)te.1943-5436.0000797.
15. Al-Kaisy, A. F., and F. Hall. Guidelines for Estimating Freeway Capacity at Long-Term Reconstruction Zones. *American Society of Civil Engineers: Journal of Transportation Engineering*, Vol. 129, No. 5, 2002, pp. 572–577.
16. Krammes, R. A., and G. Lopez. Updated Capacity Values for Short-Term Freeway Work Zone Lane Closures. *Transportation Research Record: Journal of the Transportation Research Board*, 1994. 1442: 49–56.
17. Chitturi, M., R. Benekohal, and A. Kaja-Mohideen. Methodology for Computing Delay and Users Costs in Work Zones. *Transportation Research Record: Journal of the Transportation Research Board*, 2008. 2055: 31–38.
18. Schroeder, B., and N. Rouphail. *Corridor-Based Forecasts of Work-Zone Impacts for Freeways and Multilane Highways*. Project HWY 10-08 Final Report. North Carolina Department of Transportation, Raleigh, NC, 2011.
19. Aghdashi, S., N. M. Rouphail, A. Hajbabaie, and B. J. Schroeder. Generic Speed–Flow Models for Basic Freeway Segments on General-Purpose and Managed Lanes in Undersaturated Flow Conditions. *Transportation Research Record: Journal of the Transportation Research Board*, 2015. 2483: 102–110.
20. Chase, R. T., A. Hajbabaie, and B. J. Schroeder. Incorporating Weather Effects in HCM Reliability Analysis. Presented at 92nd Annual Meeting of the Transportation Research Board, Washington D.C., 2013.
21. Rakha, H., M. Farzaneh, M. Arafeh, and E. Sterzin. Inclement Weather Impacts on Freeway Traffic Stream Behavior. *Transportation Research Record: Journal of the Transportation Research Board*, 2008. 2071: 8–18.

22. Shi, L., Y. Cheng, J. Jin, B. Ran, and X. Chen. Effects of Rainfall and Environmental Factors on Traffic Flow Characteristics on Urban Freeway. Presented at 90th Annual Meeting of the Transportation Research Board, Washington, D.C., 2011.
23. Dehman, A. Effect of Inclement Weather on Two Capacity Flows at Recurring Freeway Bottlenecks. *Transportation Research Record: Journal of the Transportation Research Board*, 2012. 2286: 84–93.
24. Aghdashi, S., N. Rouphail, and A. Hajbabaie. Estimation of Incident Propensity for Reliability Analysis in the Highway Capacity Manual. *Transportation Research Record: Journal of the Transportation Research Board*, 2013. 2395: 123–131.
25. Skabardonis, A., K. F. Petty, R. L. Bertini, P. P. Varaiya, N. H, and D. Rydzewski. The I-880 Field Experiment: Analysis of the Incident Data. *Transportation Research Record: Journal of the Transportation Research Board*, 1997. 1603: 72–79.
26. Schroeder, B., N. Rouphail, and S. Aghdashi. Deterministic Framework and Methodology for Evaluating Travel Time Reliability on Freeway Facilities. *Transportation Research Record: Journal of the Transportation Research Board*, 2013. 2396: 61–70.
27. Williges, C., B. McCullough, Y. Chu, N. Amatya, R. Kuo, M. Lin, and L. Chu. *SHRP 2 Reliability Project L38A: Pilot Testing of SHRP 2 Reliability Data and Analytical Products: Southern California*. Transportation Research Board of the National Academies, Washington, D.C., 2014.
28. Sobolewski, M., T. Polum, P. Morris, R. Loos, and K. Anderson. *SHRP 2 Reliability Project L38B: Pilot Testing of SHRP 2 Reliability Data and Analytical Products: Minnesota*. Transportation Research Board of the National Academies, Washington, D.C., 2014.
29. Hadi, M.Y. Xiao, T. Wang, P. Hu, J. Jia, R. Edelstein, and A. Lopez. *SHRP 2 Report: Pilot Testing of SHRP 2 Reliability Data and Analytical Products: Florida Pilot Site*. Transportation Research Board of the National Academies, Washington, D.C., 2014.
30. Nisbet, J., D. Bremmer, S. Yan, D. Murshed, Y. Wang, Y. Zou, W. Zhu, M. Dunlap, B. Wright, T. Zhu, and Y. Zhang. *SHRP 2 Reliability Project L38D: Pilot Test of SHRP 2 Reliability Data and Analytical Products: Washington*. Transportation Research Board of the National Academies, Washington, D.C., 2014.
31. Trask, J. L., A. Seyedbehzad, B. John, and M. R. Nagui. Capacity Calibration for Freeway Facilities Methodology in the HCM: A Metaheuristic Approach. Presented at 96th Annual Meeting of the Transportation Research Board, Washington D.C., 2017.
32. RITIS. *Regional Integrated Transportation Information System*. 2016. www.ritis.org.
33. Webmaster. *Traveler Information Management System*. NCDOT. https://tims.ncdot.gov/tims/. Accessed July 31, 2017.
34. Weather Underground. https://www.wunderground.com. Accessed July 31, 2017.
35. Chase, R. T., B. Williams, N. Rouphail, and S. Kim. Reporting Path Travel Times for ATIS Reliability. Presented at the INSTR 2012 Conference in Hong Kong, China, 2012.
36. *MAP-21 Proposed Measures for Congestion, Reliability, and Freight: Step-by-Step Calculation Procedures*. FHWA, 2016. https://www.apta.com/gap/fedreg/Documents/MAP-21_Proposed_Measures_for_Congestion%2C_Reliability%2C_and_Freight.pdf.

The Standing Committee on Highway Capacity and Quality of Service (AHB40) peer-reviewed this paper (18-06088).

The views and opinions expressed in this paper do not necessarily reflect the views of the agencies, and no official endorsement should be inferred.

Article

TRR
JOURNAL OF THE TRANSPORTATION RESEARCH BOARD

Transportation Research Record
2018, Vol. 2672(15) 105–117

DOI: 10.1177/0361198118794538
journals.sagepub.com/home/trr
SAGE

Modeling Free-flow Speeds on Undivided Roads in Mixed Traffic with Weak Lane Discipline

Vipin Chathoth[1] and Gowri Asaithambi[1]

Abstract

In developing countries like India, transportation systems are characterized by limited roadway infrastructure and lack of operation and management experience. Hence, there exists a need to evaluate a performance indicator that reflects the current level of service (LOS) of a road facility. Free-flow speed (FFS) is a key parameter used to express LOS assessment. The objective of this study is to develop FFS prediction models for undivided roads with mixed traffic conditions in both urban and rural settings in India. Traffic data were collected from two-way two-lane undivided roads in southern India during free-flow traffic conditions using videographic method. Various class-specific and site-specific characteristics, such as vehicle class, subclass, carriageway width, link length, number of side roads, lateral clearance, land use type, and area type, were investigated and their influence on FFS evaluated. Statistical tests assessed the variations of obtained FFS with different vehicle-specific and site-specific factors. Free-flow prediction models were developed using linear regression method. The developed models show that FFS increases with greater carriageway width, lateral clearance, and link length, and decreases with increase in number of side roads. In general, FFS is higher in rural areas than urban areas. Similarly, open areas have higher FFS than residential, institutional, and commercial areas. The model can be used to predict FFS of undivided roads if site-specific and vehicle-specific data are known. This study finds interesting applications in capacity and LOS analysis, accident analysis, and before-and-after studies of road improvement schemes.

Free-flow speed (FFS) can be defined as the desired average speed adopted by a driver when not restricted by other vehicles in the traffic stream under a given set of road conditions. FFS is an important parameter that should be measured accurately in the field because it plays a major role in planning, operational analysis, and performance evaluation of transportation systems. However, field determination of FFS is resource intensive because it requires measuring a sufficient sample size of vehicle speeds at the appropriate time of low vehicle interaction and in appropriate locations with comparable geometric features. To overcome the challenges of field data collection, most highway agencies use modeling techniques to predict FFS.

In developing countries like India, traffic is composed of a wide variety of different types of vehicles, such as cars, buses, trucks, light commercial vehicles, two-wheelers, three-wheelers, etc. These vehicles generally do not follow lane discipline and they occupy any position in the available road space. This type of traffic is known as mixed or heterogeneous traffic conditions. It is believed that different classes of vehicles have different speed values based on their size and characteristics. FFS also depends on site-specific characteristics such as land use, area type, roadway geometry, etc. In a literature review, it was found that the majority of studies on FFS are based on homogeneous traffic conditions. Studies on FFS under mixed traffic conditions are limited and these studies have been done mostly on divided roads. Also, not much attention has been given to site-specific characteristics in earlier studies. Hence, the main objective of the present study is to develop FFS prediction models for undivided roads under mixed traffic conditions with the following specific objectives:

- To analyze whether FFS are statistically similar for different vehicle types and site-specific factors.
- To study the influence on FFS of vehicle-specific and site-specific characteristics.

[1]National Institute of Technology Karnataka, Surathkal, Mangalore, India

Corresponding Author:
Address correspondence to Gowri Asaithambi: gowri_iitm@yahoo.co.in

- To develop overall and class-wise FFS prediction models for undivided roads under non-lane based mixed traffic conditions.

The rest of the paper is structured as follows: the next section focuses on a detailed review of the literature on FFS studies. The third section discusses the data collection and extraction process. The fourth section presents the results of analysis of FFS based on vehicle-specific and site-specific characteristics. FFS prediction models are presented in the fifth section, followed by a summary and conclusions.

Literature Review

FFS is defined as the average speed of a vehicle on a given facility, measured under low-volume conditions, when drivers tend to drive at their desired speed and are not constrained by control delay. The definitions of FFS given by the *Highway Capacity Manual* 2010 (*1*) are:

- The theoretical speed when density and flow rate on a study segment are both zero.
- The prevailing speed on freeways at flow rates of between 0 and 1,000 passenger cars per hour per lane (pc/h/ln)

Bang (*2*) studied the effect of side friction factors like carriage width, presence of shoulder or kerbs, shoulder width, side friction events, roadside land use, etc. on FFS. It was found that the impact of side friction was severe on FFS for both two-lane undivided roads, both interurban and urban.

Dixon et al. (*3*) quantified the relationship between posted speed limit and observed FFS on rural multilane highways. Some of the issues evaluated in that study include influence on FFS of: heavy vehicles, traffic volumes, access point density, and vertical grade. Ye et al. (*4*) also explored the effect of truck percentage, speed limit (65 and 55 miles/h), land use type (rural/urban), road classification (freeway/non-freeway), and number of lanes (four/six). The results show that vehicles in rural areas had higher FFS than in urban areas for all models, and vehicles on freeways had higher speed than on non-freeways. Kyte et al. (*5*) studied the effect of environmental factors on FFS. Data was collected using sensors which measured traffic, visibility, roadway, and weather data. The study concluded that wind speed, low lighting, and snow reduce driver speed. Hablas (*6*) attempted to quantify the impact of inclement weather (precipitation and visibility) on FFS of traffic stream along freeway sections. Wang (*7*) conducted a study to examine the relationship between FFS and posted speed limits on expressways in Beijing, China. A regression model was built to establish a relationship between 85th percentile speed and posted speed limit. Hashim (*8*) investigated the relationship between 85th percentile speed and headway to define a headway value corresponding to free moving vehicles. He found that the 85th percentile speed took a constant value at headway equal to five seconds or more. Azai and Vien (*9*) studied the factors affecting FFS on multilane highways in Malaysia. Further investigation on lane width, shoulder width, and access point was also done. Moses and Mtoi (*10*) developed FFS equations for urban arterials in Tallahassee, Florida, and compared the efficacy of models with the HCM 2010 (*1*) and Florida Department of Transportation methods. Wang et al. (*11*) developed a linear regression model to estimate FFS on freeways in California in order to investigate the impact of changes in pavement roughness on driving behavior with respect to speed. Andrade et al. (*12*) used multivariate analysis techniques to investigate the effect on FFS of posted speed, road class, number of lanes, access point density, geometric design, and roadside environment. Ma et al. (*13*) found that variance in FFS across study sections increased with increase in lane width at midblock, but did not show any significant impact at the exits of intersections. Tunde (*14*) used regression analysis to find the relationship between various factors (driver, vehicle, road geometry, and environment) and FFS on urban arterials in Nigeria. Tseng et al. (*15*) investigated the characteristics of FFS in relation to vehicle type, geometric design, and speed limit. Puan and Abdurrahman (*16*) made a comparison of FFS estimates obtained using moving car observer method with the Malaysian *Highway Capacity Manual*. Balakrishnan and Sivanandan (*17*) studied the influence on FFS of lane and vehicle subclass for urban roads in mixed traffic. Analysis of FFS was performed for all the vehicle-specific and site-specific factors. FFS prediction models were developed using multiple linear regression method, and multinomial logistic regression was employed for developing lane choice models. Rao and Rao (*18*) developed a model for FFS obtained in urban arterials of Delhi and compared it with the FFS obtained from developed and HCM 2010 (*1*) models. They found that the FFS value given by HCM 2010 (*1*) is higher than the field value. Yusuf et al. (*19*) performed statistical analysis and found that the environment, pedestrians, and roadway geometry have negative influences on FFS.

Speed and travel time studies are necessary for the determination of road performance. The literature review confirms that researchers have conducted studies on FFS to understand driving behavior and factors affecting it. These factors are generally related to drivers, roadway geometry, and land use. It is notable that the majority of the studies on FFS are based on homogeneous traffic

Table 1. Site-Specific Details of Study Sections

Site no.	Site	Link length (m)	Distance between cameras (m)	CW (m)	Side clearance (m)	No. of side roads	No. of vehicles	Area type	Land use
1	Padubidri	693.3	320.3	8.6	2.4	2	505	Suburban	Residential
2	Father Muller	788.7	249.0	8.2	1.5	4	414	Urban	Residential
3	Yenapoya	770.6	198.8	7.0	1.0	2	781	Urban	Institutional
4	Vamanjoor	669.0	326.5	5.5	3.0	2	821	Rural	Open
5	AC Road	1700.0	300.0	9.7	1.3	0	1094	Rural	Open
6	Convent Square	717.0	271.1	6.5	0.2	6	511	Suburban	Commercial
7	Thumpoli	764.4	336.6	10.0	1.5	4	1905	Urban	Commercial
8	Chovva	1300.0	305.8	6.8	1.6	2	1098	Suburban	Open
9	Puthiyatheru	960.0	185.3	10.0	1.5	2	1257	Rural	Open
10	Nadal	1540.0	401.3	7.0	1.0	0	1214	Rural	Open

Note: CW = carriageway width.

conditions. Studies on FFS under mixed traffic conditions are limited in number. It was found that most of the factors influencing FFS in homogeneous traffic are applicable to mixed traffic also. However, it is evident that the presence of various vehicle groups is an important determining factor in mixed traffic conditions. Studies available for undivided road stretches with mixed traffic conditions are limited. Hence, it is necessary to develop a model for predicting the FFS of undivided roads, which are predominant in most of the rural and urban road network of India.

Data Collection and Extraction

In order to study and model the FFS on undivided roads, ten study sections were selected on various road stretches located in southern cities of India: Mangalore, Kannur, and Alappuzha. Traffic data were collected using videographic technique from all the chosen sites. The collected data were classified into following two groups:

1. Vehicle-specific data
   a. Vehicle class and subclass
   b. Number of vehicles (vehicles moving in both directions)
2. Site-specific data
   a. Carriageway width
   b. Link length
   c. Side clearance
   d. Number of side roads (roads which perpendicularly merge with the study stretch)
   e. Land use (residential, institutional, commercial, and open)
   f. Area type (rural, suburban, and urban)

The video data were collected from each site using a pair of camcorders for 1.5 hours which were synchronised for time. The data collection was done during the early morning hours, that is, between 5:30 am and 7:30 am, depending on the traffic flow rate, in order to observe free-flow conditions. The details of link length, geometric details, number of observations, and site-specific characteristics are shown in Table 1.

Vehicle-specific details are extracted from the collected videos using spreadsheet and media player software. The travel times of vehicles for each site were estimated using vehicle re-identification technique. The difference in entry and exit times of a particular vehicle gives its free-flow travel time. From the known distance and travel time, FFS of each vehicle was estimated. Average FFS of vehicles in both directions is taken as the FFS since undivided roads are the subject of the study. The individual speeds are then aggregated into class-wise speeds and overall (all vehicle classes together) speeds for five-minute intervals. In the present study, vehicles are classified into six categories and the posted speed limits of each study site are shown in Table 2.

Analysis of FFS

In this section, the details of vehicle composition in each site are given. The cumulative percentage of speeds based on vehicle-specific and site-specific characteristics are presented in this section. The influences of site-specific and vehicle-specific factors on FFS are also examined.

Vehicular Composition

In the present study, a total of 9,600 vehicles were analyzed, of which 34% were two-wheelers, 8% three-wheelers, 30% cars, and 7% light commercial vehicles, 7% buses, and 14% trucks. The majority of the vehicles were

Table 2. Details of Vehicle Classes and Subclasses Considered Including Speed Limits for Different Sites

Sl. no.	Class	Notation	Posted speed limits of sites (km/h)										Subclass
			1	2	3	4	5	6	7	8	9	10	
1	Two-wheelers	TW	50	50	50	60	50	50	50	60	60	60	Motor bikes & Scooters
2	Three-wheelers	ThW	50	30	30	50	50	40	30	50	50	50	Passenger carriers & goods carriers
3	Passenger cars	Car	80	50	50	85	80	70	50	85	85	85	Hatchbacks, sedans & SUVs
4	Light commercial vehicles	LCV	65	50	50	65	65	60	50	65	65	65	Passenger carriers & goods carriers
5	Buses	Bus	65	40	40	65	65	60	40	65	65	65	Passenger carriers & institutional buses
6	Trucks	Truck	65	40	40	65	65	60	40	65	65	65	Trucks

Note: SUV = sports utility vehicle.

either two-wheelers or passenger cars. The proportion of light commercial vehicles and buses was found to be the minimum. Figure 1 shows the vehicle composition of the selected ten sections.

Cumulative Percentile Speed Curves

The 85th and 15th percentile speeds give a general description of the high and low speeds observed by most reasonable drivers. The 85th percentile speed is the speed that 85 percent of vehicles do not exceed. The speed limit is commonly set at or below the 85th percentile operating speed. The 50th percentile speed represents the average speed of the traffic stream. Figure 2 represents cumulative percentage frequency distribution curves for speeds of traffic stream (all vehicles together), different class of vehicles, area type, and land use types.

The overall 85th percentile speed obtained is 58 km/h. The 50th percentile speed and 15th percentile speeds obtained are 43 km/h and 32 km/h, respectively. Cars have the highest 85th percentile speed (64 km/h) and three-wheelers have the lowest (43 km/h). Considering area types, suburban areas have higher 85th percentile speed (60 km/h) followed by rural (57 km/h) and urban (55 km/h) areas. The 50th percentile speed also follows the same trend as the 85th percentile speed but the 15th percentile speed is higher in rural areas (34 km/h). Open area land use type has higher 85th, 50th, and 15th percentile speeds. The 85th percentile speeds of institutional, commercial, residential, and open areas are 53 km/h, 55 km/h, 57 km/h, and 58 km/h, respectively. Similarly, the 15th percentile speeds are 28 km/h, 29 km/h, 31 km/h, and 4 km/h for residential, commercial, institutional, and open areas, respectively. The 50th percentile speed is

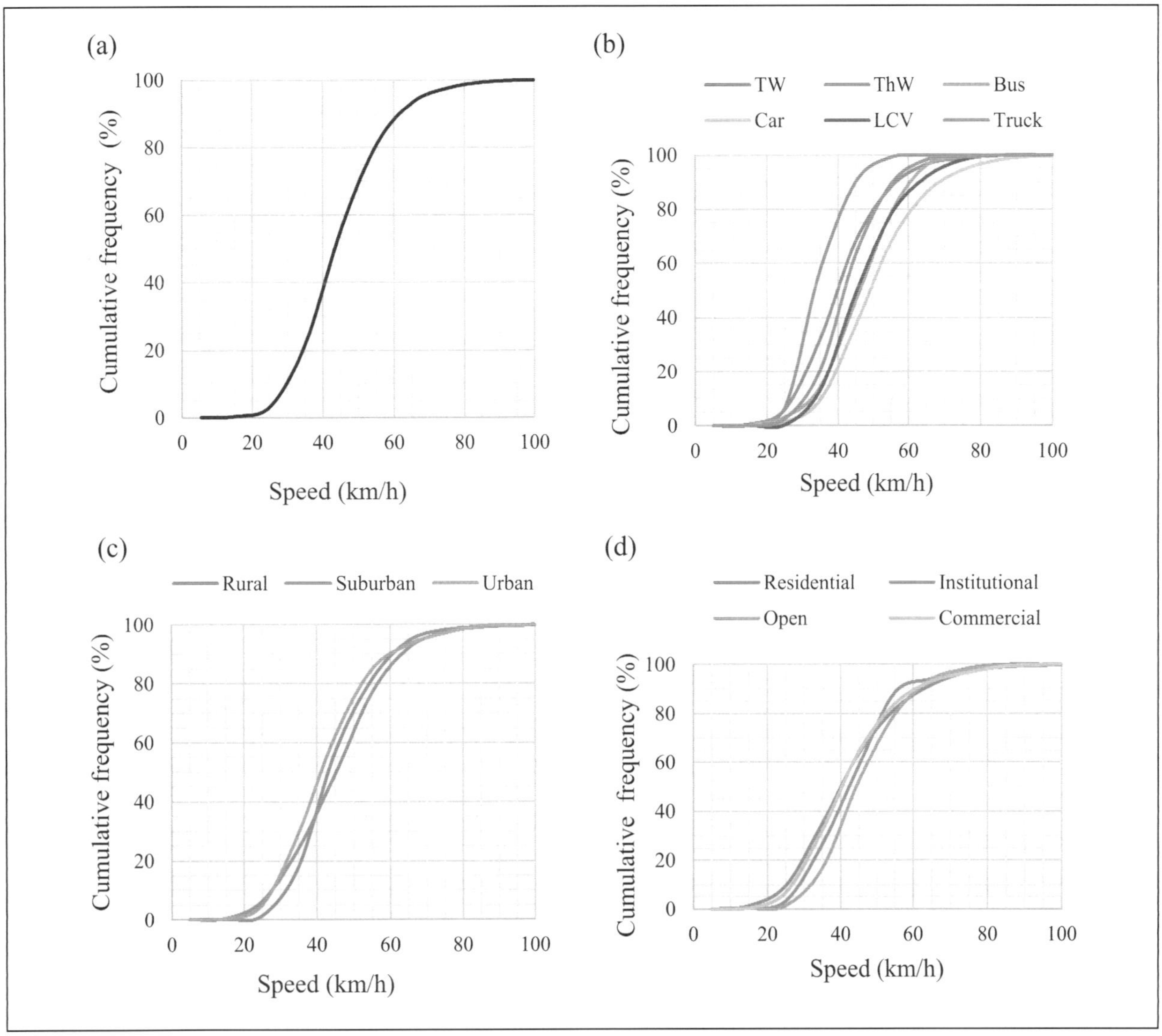

Figure 2. Cumulative percentile speed: (*a*) traffic stream; (*b*) class-wise; (*c*) area type; (*d*) land use.

highest in open areas (44 km/h) and lowest in residential areas (40 km/h).

Analysis of FFS for Different Vehicle Classes and Subclasses

The mean and the median FFS of vehicle classes and subclasses were almost equal, indicating normal distribution. Among vehicle classes, passenger cars (53.7 km/h) were found to be the fastest and three-wheelers (39.1 km/h) the slowest. Among subclasses considered in the study, the highest average speed was for sports utility vehicles (SUVs) (57.3 km/h) and the lowest for three-wheeler goods carriers (37.8 km/h). Considerable variation in standard deviation of speeds was also observed among vehicle groups.

From Table 3 it is clear that, among two-wheelers, motorbikes and scooters had statistically different mean FFS. Within three-wheelers, passenger carriers had higher FFS than goods carriers. Among cars, the variance was found to be same for all subclasses. The FFS of SUVs was found to be the highest and that of hatchback the lowest. In light commercial vehicles, passenger carriers had higher FFS and variance than goods carriers. Among buses the FFS of both subclasses was found to be similar.

Analysis of FFS for Different Area Type

Analysis was performed to understand the speed distributions of vehicle classes on urban, suburban, and rural roads. Of the ten study sections, three were chosen from

Table 3. Statistical Tests To Compare Distributions of FFS for Different Subclasses

Class	Subclass pair tested (a) – (b)	Mean FFS (km/h) (a)	Mean FFS (km/h) (b)	F-test F-stat	F-test p-value	t-test t-stat	t-test p-value
TW	Scooter – motor bike	43.0	45.0	1.179	0.137	−3.236	0.001
ThW	Goods cr. – passenger cr.	38.0	40.0	0.953	0.417	2.742	0.007
CAR	Sedan – hatchback	56.0	53.0	1.027	0.433	2.525	0.012
	SUV – hatchback	57.5	53.0	1.023	0.442	3.651	0.000
	Sedan – SUV	56.0	57.5	1.004	0.491	−1.030	0.304
LCV	Passenger cr. – goods cr.	52.0	51.0	1.295	0.999	1.970	0.501
Bus	Passenger cr. – institutional	50.5	49.0	1.405	0.057	1.009	0.316
Truck	NA	NA	NA	NA	NA	NA	NA

urban areas, three from suburban areas, and four from rural areas. The overall and class-wise FFS variations are given in box plots in Figure 3*a* and 3*c*.

The mean overall FFS on rural roads, suburban roads, and urban roads was found to be 47.6 km/h, 46.4 km/h, and 43.4 km/h, respectively. On urban roads speeds are generally lower due to side friction. The statistical significance of the difference in means and variances of FFS on different area types were checked using F-tests and t-tests (Table 4). FFS is lowest for urban areas for all vehicle classes. Also, cars were found to have the highest FFS whereas three-wheelers have the lowest in all the area types.

Analysis of FFS for Different Land Use Types

Land use represents the roadside activity and its intensity. FFS were grouped and analyzed based on various land use types. The highest overall FFS was observed in open areas, with a mean of 49.0 km/h, while the speed was lowest in commercial areas with a mean of 43.5 km/h. The mean FFS in institutional areas and residential areas was 46.1 km/h and 43.8 km/h, respectively. The overall and class-wise FFS variations are given in Figure 3*b* and *d*.

Statistical tests were conducted to understand the significance of difference in FFS across land use types for each vehicle class and the results are shown in Table 5. For all vehicle types, highest FFS was observed in open areas, followed by institutional areas, residential areas, and commercial areas. The reason for higher speed in open areas is the reduced side friction. Speeds in commercial and residential areas are lower due to buildings, more number of side roads, less lateral clearance, etc.

Effect of Site-Specific Factors on FFS

The effect of quantitative variables on FFS was analyzed, and graphs with trend lines were plotted (Figure 4). The graphs show that FFS increases with increase in link length, carriageway width, and lateral clearance and decreases with increase in number of side roads.

Models for FFS

The FFS models were developed using regression analysis in SPSS software. Two sets of models were developed: general models and class-wise models. The general models were developed considering FFS of all vehicles classes together. The class-wise models include separate models for each of the six vehicle classes considered in the study. The foremost assumptions of regression analysis are that there must be no correlation among independent variables. Though this assumption is not perfectly attainable, minimizing the occurrence of correlated variables must be ensured in variable selection. Variable combinations are selected with the help of correlation matrix.

HCM Method of FFS Estimation

HCM 2010 (*1*) suggests the following equation to estimate FFS:

$$FFS = BFFS - f_{LS} - f_A \quad (1)$$

where FFS = estimated FFS (km/h); $BFFS$ = base FFS (km/h); f_{LS} = adjustment for lane and shoulder widths; f_A = adjustment for access points.

The HCM method is valid only for homogeneous traffic and, in this study, other factors specific to Indian traffic conditions are also considered to develop FFS models.

General Model

FFS is selected as the dependent variable and total number of vehicles, heavy vehicle percentage, link length, carriageway width, lateral clearance, number of side roads, area type, and land use are selected as independent variables. Three models, GM1, GM2, GM3, are selected based on model goodness of fit, significance of model,

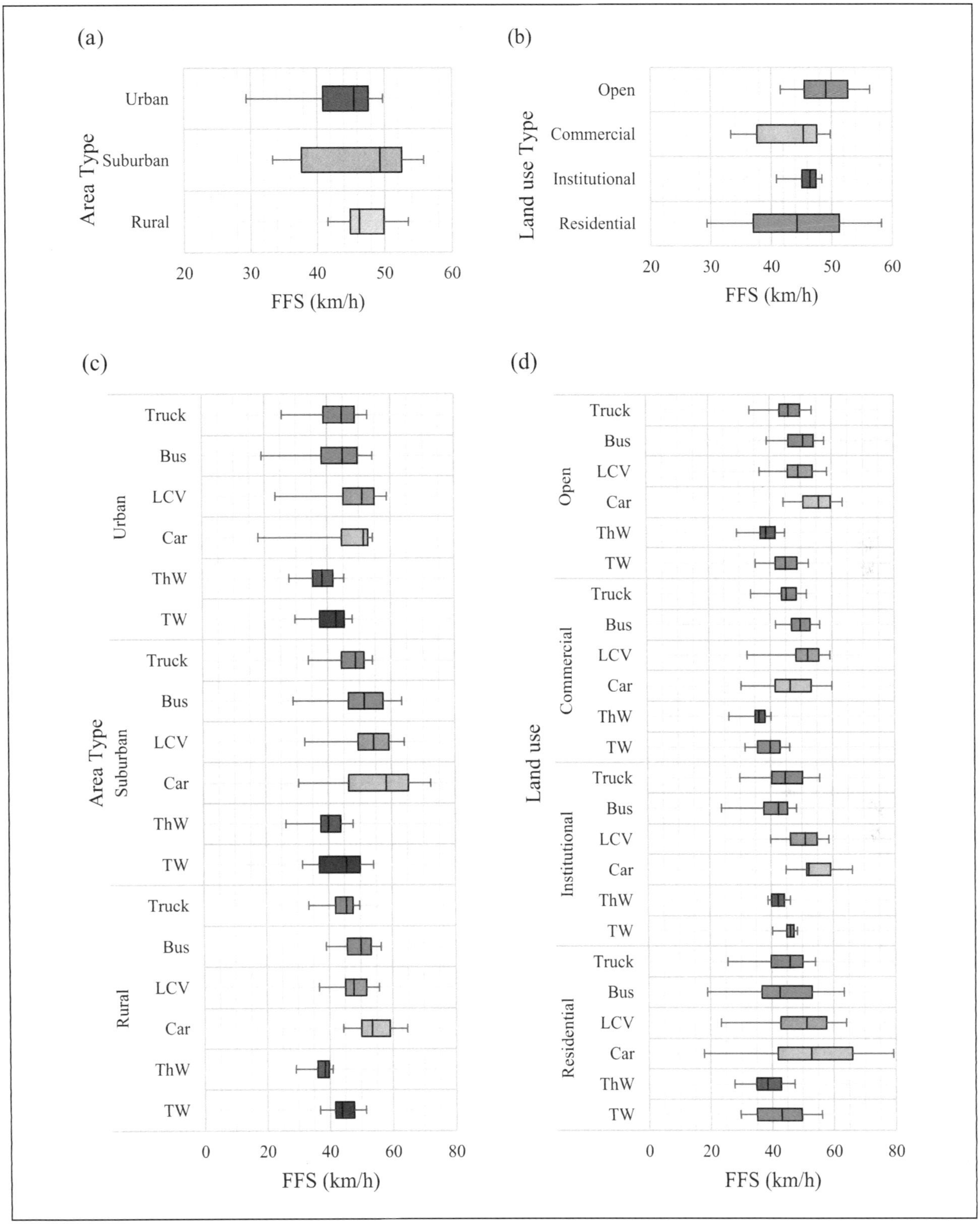

Figure 3. Box plots showing variation of FFS based on area type and land use type: (*a*) FFS on different area types; (*b*) FFS on different land use types; (*c*) class-wise FFS based on area type; (*d*) class-wise FFS based on land use type.

Table 4. Statistical Tests To Compare Distributions of FFS for Different Area Types

Class	Area type pairs tested (a) – (b)	Mean FFS (km/h) (a)	Mean FFS (km/h) (b)	F-value F-stat	F-value p-value	t-value t-stat	t-value p-value
TW	Suburban – rural	45.0	45.0	3.192	0.000	2.577	0.011
	Urban – rural	41.5	42.0	1.627	0.028	0.161	0.872
	Suburban – urban	45.0	41.5	1.962	0.008	–3.462	0.001
ThW	Suburban – rural	41.0	38.5	2.304	0.001	2.605	0.011
	Urban – rural	39.0	38.5	1.811	0.011	0.117	0.907
	Suburban – urban	41.0	39.0	1.272	0.194	2.154	0.034
Car	Suburban – rural	56.5	55.5	2.768	0.000	0.551	0.583
	Urban – rural	49.0	55.5	1.455	0.070	–4.369	0.000
	Suburban – urban	56.5	49.0	1.903	0.011	0.551	0.583
LCV	Suburban – rural	54.5	49.5	1.387	0.113	3.146	0.002
	Urban – rural	49.5	49.5	1.374	0.115	–0.078	0.938
	Suburban – urban	54.5	49.5	1.010	0.485	2.809	0.006
Bus	Suburban – rural	51.5	50.5	1.704	0.030	0.551	0.584
	Urban – rural	44.0	50.5	1.596	0.040	–4.228	0.000
	Suburban – urban	51.5	44.0	1.068	0.414	3.744	0.000
Truck	Suburban – rural	49.0	45.5	2.008	0.004	2.818	0.006
	Urban – rural	43.5	45.5	1.705	0.025	–1.653	0.103
	Suburban – urban	49.0	43.5	1.178	0.300	3.650	0.000

and significance of independent variables (Table 6). The R^2 values of these models range from 0.299 to 0.771. Collinearity test, normality test, heteroskedasticity test, etc. are also done to verify the validity of the model. Among the general models, GM3 is recommended for having higher R^2 value (0.771).

Model GM1 shows that FFS will be reduced by 0.036 km/h for an increase of one vehicle. For 1 m increase in carriage width and lateral clearance, FFS increases by 1.274 km/h and 1.683 km/h, respectively and increase in side roads decreases the speed by 1.147 km/h. Model GM2 predicts that for every 1 m increase in carriageway width, FFS increases by 1.334 km/h and for every 1 m increase in lateral clearance, speed increases by 2.836 km/h. FFS in suburban areas is 0.802 km/h lower than in rural areas. Moreover, in urban areas it is 3.423 km/h lower.

In model GM3, a 1 m increase in carriage width and lateral clearance increases FFS by 2.004 km/h and 4.041 km/h, respectively. Similarly, for one-unit increase in number of vehicles and number of side roads, FFS decreases by 0.048 km/h and 2.595 km/h, respectively. Also, FFS for land use types such as residential areas, institutional areas, and commercial areas are 9.751 km/h, 3.925 km/h and 9.476 km/h respectively, which are lower than for open area land use.

With reference to the model, carriageway width and lateral clearance have a positive effect on FFS. This is logical, as increase in carriageway width and lateral clearance enhances the smooth flow of traffic. Similarly, it can be inferred from the above model that the number of vehicles and number of side roads has a negative effect on FFS. That is, as the number of vehicles and number of side roads increases, it interrupts the free-flow of traffic.

Class-Wise Model

To study the effect of various factors on different vehicle classes, a class-wise regression model is developed for six classes of vehicles. Tables 7 and 8 show the goodness of fit, significance of model, and significance of independent variables for class-wise models of specifications 1 and 2.

FFS from specification 1 model can be obtained using Equation 1.

$$\begin{aligned} FFS(x) = b_0 &+ (b_1 \times No.of\ Vehicles) \\ &+ (b_2 \times Carriageway\ Width) \\ &+ (b_3 \times Lateral\ Clearance) \\ &+ (b_4 \times No.of\ Side\ Roads) \\ &+ (b_5 \times Area\ Suburban) \\ &+ (b_6 \times Area\ Urban) \end{aligned} \tag{2}$$

Similarly, FFS from specification 2 model is obtained using Equation 2.

$$\begin{aligned} FFS(x) = b_0 &+ (b_1 \times Carriageway\ Width) \\ &+ (b_2 \times Lateral\ Clearance) \\ &+ (b_3 \times Area\ Suburban) \\ &+ (b_4 \times Land\ use\ Residential) \\ &+ (b_5 \times Land\ use\ Institutional) \\ &+ (b_6 \times Land\ Use\ Open) \end{aligned} \tag{3}$$

Table 5. Statistical Tests To Compare Distributions of FFS for Different Land Use Types

Class	Land use pairs tested	Mean FFS (km/h)		F-value		t-value	
	(a) – (b)	(a)	(b)	F-stat	p-value	t-stat	p-value
TW	Residential – institutional	43.0	46.5	8.611	0.000	–2.379	0.021
	Residential – commercial	43.0	39.5	3.907	0.000	2.059	0.044
	Residential – open	43.0	45.5	2.987	0.000	–1.842	0.072
	Commercial – institutional	39.5	46.5	2.204	0.042	–7.083	0.000
	Open – institutional	45.5	46.5	2.883	0.008	–1.123	0.268
	Open – commercial	45.5	39.5	1.308	0.188	6.837	0.000
ThW	Residential – institutional	39.5	42.5	4.337	0.001	–2.661	0.010
	Residential – commercial	39.5	37.0	2.413	0.005	2.405	0.019
	Residential – open	39.5	39.5	1.995	0.005	–0.009	0.993
	Commercial – institutional	37.0	42.5	1.798	0.105	–6.428	0.000
	Open – institutional	39.5	42.5	2.175	0.041	–3.881	0.001
	Open – commercial	39.5	37.0	1.210	0.268	3.640	0.001
Car	Residential – institutional	54.0	55.0	6.565	0.000	–0.319	0.751
	Residential – commercial	54.0	47.5	3.143	0.001	2.271	0.027
	Residential – open	54.0	56.0	4.990	0.000	–0.870	0.389
	Commercial – institutional	47.5	55.0	2.089	0.054	–3.749	0.000
	Open – institutional	56.0	55.0	1.316	0.267	0.871	0.391
	Commercial – open	47.5	56.0	1.588	0.044	5.516	0.000
LCV	Residential – institutional	50.0	51.0	2.778	0.018	–0.436	0.665
	Residential – commercial	50.0	51.5	2.292	0.018	–0.649	0.520
	Residential – open	50.0	51.0	2.363	0.002	–0.386	0.702
	Commercial – institutional	51.5	51.0	1.212	0.350	0.228	0.821
	Open – institutional	51.0	51.0	1.176	0.374	–0.163	0.872
	Commercial – open	51.5	51.0	1.031	0.439	0.489	0.627
Bus	Residential – institutional	43.5	42.0	3.155	0.008	0.683	0.498
	Residential – commercial	43.5	50.0	6.030	0.000	–2.686	0.011
	Residential – open	43.5	51.5	2.242	0.003	–3.374	0.002
	Institutional – commercial	42.0	50.0	1.911	0.091	–4.525	0.000
	Open – institutional	51.5	42.0	1.407	0.216	5.658	0.000
	Open – commercial	51.5	50.0	2.689	0.010	1.002	0.322
Truck	Residential – institutional	45.0	44.5	1.398	0.268	0.141	0.889
	Commercial – residential	46.5	45.0	1.157	0.358	0.848	0.400
	Residential – open	45.0	46.5	2.394	0.001	–1.082	0.288
	Commercial – Institutional	46.5	44.5	1.618	0.181	0.927	0.361
	Institutional – open	44.5	46.5	1.712	0.072	–1.131	0.276
	Commercial – open	46.5	46.5	2.771	0.000	0.052	0.959

where x is the class of vehicle.

R^2 values of these models range from 0.305 to 0.649. CM2.TW model is recommended based on higher R^2 value (0.649). It can be concluded from all the models that both carriageway width and lateral clearance have a positive effect on FFS. That is, as carriageway width and lateral clearance increases, FFS also increases. From specification 1, it is seen that FFS decreases with increase in number of vehicles in the traffic stream, except for heavy vehicles (buses and trucks). Also, it was found that as the number of side roads increases FFS decreases. The speed of vehicles in urban and suburban areas is found to be lower than in rural areas. It is also seen that for different land use types, different vehicles have different trends in FFS.

Prediction of Overall FFS using Class-Wise Models

In the present study, class-wise models are used to estimate the FFS of six classes of vehicles. These models can be employed to obtain the overall FFS, as shown in Equation 3.

$$\begin{aligned} FFS = b_0 &+ (P_{TW} \times FFS_{TW}) + (P_{ThW} \times FFS_{ThW}) \\ &+ (P_{Car} \times FFS_{Car}) + (P_{LCV} \times FFS_{LCV}) \\ &+ (P_{Bus} \times FFS_{Bus}) + (P_{Truck} \times FFS_{Truck}) \end{aligned} \quad (4)$$

where

FFS is the overall FFS;

FFS_{TW}, FFS_{ThW}, ..., FFS_{Truck} are the class-wise FFS in the study stretch

P_{TW}, P_{ThW}, ..., P_{Truck} are the class-wise proportions.

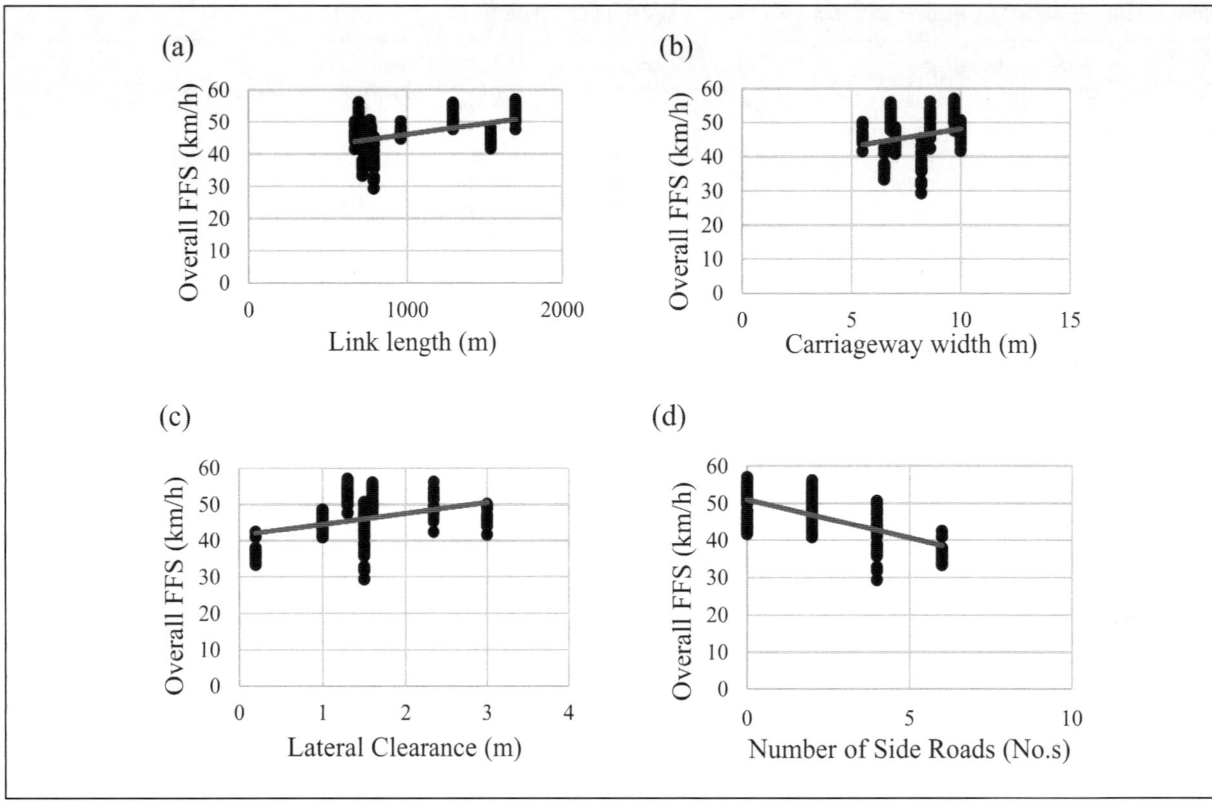

Figure 4. Effect of various site-specific factors on overall FFS: (*a*) link length; (*b*) carriageway width; (*c*) lateral clearance; (*d*) number of side roads.

Table 6. General Models: Estimation of Coefficients, Goodness of Fit, and Significance

Model	*R* square	Variables	Coefficients	Std. error	*t*-value	Sig.
GM1	0.299	(Constant)	35.083	3.324	10.553	0.000
		No. of vehicles	−0.036	0.387	2.346	0.001
		Carriageway width	1.274	0.294	4.332	0.000
		Lateral clearance	1.683	0.531	3.169	0.002
		No. of side roads	−1.147	0.322	−3.567	0.001
GM2	0.503	(Constant)	32.228	2.773	11.621	0.000
		Carriageway width	1.334	0.297	4.491	0.000
		Lateral clearance	2.836	0.617	4.596	0.000
		Area suburban	−0.802	1.087	0.738	0.462
		Area urban	−3.423	1.070	−3.198	0.002
GM3	0.771	(Constant)	40.051	1.812	22.097	0.000
		No. of vehicles	−0.048	0.016	−3.029	0.003
		Carriageway width	2.004	0.240	8.354	0.000
		Lateral clearance	4.041	0.451	8.956	0.000
		No. of side roads	−2.595	0.195	−13.302	0.000
		Land use residential	−9.751	0.849	−11.485	0.000
		Land use institutional	−3.925	0.692	−5.676	0.000
		Land use commercial	−9.476	1.169	−8.106	0.000

Validation of Models

In the validation stage, an attempt was made to understand how well the models could predict the outcome of an independent sample from the same population. The FFS prediction models were built using 72.2% of data collected. The models were internally validated using the

Table 7. Class-wise Models, Specification 1: Coefficients, Goodness of Fit, and Significance

Model	*R* square	Variables	B	Std. error	*t*-value	Sig.
CM1.TW	0.623	(Constant)	41.21	2.388	17.254	0.000
		No. of vehicles	−0.059	0.021	−2.829	0.005
		Carriageway width	1.32	0.316	4.176	0.000
		Lateral clearance	5.075	0.595	8.534	0.000
		No. of side roads	−2.297	0.257	−8.932	0.000
		Area suburban	−3.161	0.911	−3.468	0.001
		Area urban	−10.202	1.119	−9.118	0.000
CM1.ThW	0.347	(Constant)	39.184	2.523	15.528	0.000
		No. of vehicles	−0.105	0.022	−4.785	0.000
		Carriageway width	1.183	0.332	3.559	0.001
		Lateral clearance	2.596	0.626	4.149	0.000
		No. of side roads	−1.27	0.272	−4.678	0.000
		Area suburban	−1.121	0.959	−1.168	0.245
		Area urban	−7.066	1.187	−5.954	0.000
CM1.CAR	0.631	(Constant)	45.297	3.856	11.747	0.000
		No. of vehicles	−0.176	0.034	−5.187	0.000
		Carriageway width	3.443	0.51	6.745	0.000
		Lateral clearance	7.328	0.96	7.632	0.000
		No. of side roads	−3.652	0.415	−8.798	0.000
		Area suburban	−7.228	1.471	−4.912	0.000
		Area urban	−15.81	1.806	−8.753	0.000
CM1.LCV	0.324	(Constant)	39.779	4.405	9.031	0.000
		No. of vehicles	−0.081	0.039	−2.055	0.042
		Carriageway width	2.947	0.585	5.035	0.000
		Lateral clearance	2.98	1.161	2.567	0.012
		No. of side roads	−1.416	0.555	−2.551	0.012
		Area suburban	−6.297	1.856	−3.393	0.001
		Area urban	−12.046	2.114	−5.697	0.000
CM1.BUS	0.603	(Constant)	22.403	8.477	2.643	0.009
		No. of vehicles	0.156	0.074	2.098	0.038
		Carriageway width	2.725	1.122	2.428	0.017
		Lateral clearance	11.117	2.111	5.267	0.000
		No. of side roads	−6.841	0.913	−7.497	0.000
		Area suburban	−4.565	3.235	−1.411	0.161
		Area urban	−17.158	3.971	−4.321	0.000
CM1.TRUCK	0.312	(Constant)	27.281	8.625	3.163	0.002
		No. of vehicles	0.064	0.076	0.841	0.402
		Carriageway width	2.611	1.142	2.287	0.024
		Lateral clearance	3.57	2.148	1.662	0.099
		No. of side roads	−1.519	0.929	−1.636	0.104
		Area suburban	−16.492	3.291	−5.011	0.000
		Area urban	−10.93	4.04	−2.705	0.008

same 72.2% data (internal validation) and externally validated using the remaining 27.8% data (external validation). The data for model development and validation were selected randomly from the whole sample. The predicted values of FFS were compared with the observed FFS and mean absolute percentage error (MAPE) was computed. MAPE values of general and class-wise models are found to vary between 7% and 25%. MAPE values of all general and class-wise models, except class-wise model for buses and trucks, are found to vary between 7% and 20% which shows that the models are reasonably successful at predicting FFS.

Summary and Conclusions

The present study aims to develop models for predicting free-flow speed (FFS) on undivided roads, which are common road facilities in India. The study evolved the variability in FFS among different vehicle classes which are considered to be an important aspect in mixed traffic conditions. Data were collected from ten different sites having different site-specific characteristics located in southern parts of India using videographic technique. Variations of FFS with site-specific and vehicle-specific factors were checked using statistical tests and FFS

Table 8. Class-wise Models, Specification 2: Coefficients, Goodness of Fit, and Significance

Model	R square	Variables	Coefficients	Std. error	t-value	Sig.
CM2.TW	0.649	(Constant)	12.228	2.679	4.564	0.000
		Width	2.324	0.267	8.691	0.000
		Lateral clearance	4.825	0.524	9.203	0.000
		Land use residential	−1.820	1.036	−1.756	0.081
		Land use institutional	12.823	1.335	9.607	0.000
		Land use open	6.885	0.937	7.345	0.000
CM2.ThW	0.317	(Constant)	23.638	2.981	7.929	0.000
		Carriageway width	1.138	0.298	3.820	0.000
		Lateral clearance	1.545	0.582	2.656	0.009
		Land use residential	1.662	1.142	1.455	0.148
		Land use institutional	8.976	1.506	5.960	0.000
		Land use open	4.049	1.055	3.840	0.000
CM2.CAR	0.620	(Constant)	−1.264	4.551	−0.278	0.782
		Carriageway width	4.400	0.454	9.686	0.000
		Lateral clearance	6.309	0.891	7.085	0.000
		Land use residential	1.489	1.760	0.846	0.399
		Land use industrial	20.014	2.268	8.826	0.000
		Land use open	12.654	1.592	7.947	0.000
CM2.LCV	0.330	(Constant)	16.232	6.112	2.656	0.009
		Carriageway width	3.280	0.564	5.818	0.000
		Lateral clearance	2.517	1.099	2.291	0.024
		Land use residential	−2.136	2.248	−0.950	0.344
		Land use institutional	9.734	2.767	3.518	0.001
		Land use open	4.371	1.911	2.287	0.024
CM2.BUS	0.517	(Constant)	−53.299	10.874	−4.902	0.000
		Carriageway width	8.034	1.085	7.403	0.000
		Lateral clearance	13.954	2.128	6.559	0.000
		Land use residential	−15.619	4.206	−3.714	0.000
		Land use institutional	24.527	5.417	4.527	0.000
		Land use open	18.207	3.804	4.786	0.000
CM2.TRUCK	0.305	(Constant)	−3.998	10.079	−0.397	0.692
		Carriageway width	4.303	1.006	4.278	0.000
		Lateral clearance	4.347	1.972	2.204	0.029
		Land use residential	−14.295	3.898	−3.667	0.000
		Land use institutional	0.383	5.021	0.076	0.939
		Land use open	9.031	3.526	2.561	0.012

prediction models were developed. General (all vehicles together) and class-wise models were developed using SPSS software. The key conclusions arising from the study are:

- The overall 85th percentile speed, 50th percentile speed and 15th percentile speed of vehicles considering all sites are 58 km/h 43 km/h and 32 km/h, respectively.
- Cars have highest 85th percentile speed (64 km/h) and three-wheelers have the lowest (43 km/h). While considering area type, suburban have higher 85th percentile speed (60 km/h) and urban has the lowest (55 km/h). Open area land use type has the highest 85th percentile speed (58 km/h) followed by residential areas (57 km/h).
- FFS of six vehicle classes under study vary significantly. The increasing order of FFS among the vehicle classes is as follows: three-wheelers (39.1 km/h), two-wheelers (43.7 km/h), trucks (46.0 km/h), buses (48.6 km/h), light commercial vehicles (50.7 km/h) and cars (53.6 km/h).
- Mean overall FFS in rural areas (47.6 km/h) is higher than suburban (46.4 km/h) and urban areas (43.4 km/h). On urban roads speeds are generally lesser due to side friction.
- Mean overall FFS in open areas (48.9 km/h) is higher than residential, institutional, and commercial areas. The reason for higher speeds in open areas is due to reduced side friction in open areas. FFS in commercial and residential areas is lower due to buildings, more number of side roads, lesser lateral clearance, etc.
- From all the models and statistical tests, it is clear that carriageway width, lateral clearance, and link length have a positive effect on FFS.

- From general models, it can be inferred that number of vehicles and number of side roads adversely affects FFS.
- Class-wise models show that the influence of site factors on FFS varies significantly across vehicle classes.

The developed models can be used for planning and operational analysis of undivided roads. The findings will be helpful to traffic planners and engineers for regulation and control of traffic operations, geometric design of roads, accident analysis, before-and-after studies of road improvement schemes, assessing journey times, and congestion on roads and in correlating capacity with speeds.

The study findings can be generalized by collecting data from various sites and also, by selecting additional independent variables (e.g., driver behavior, vehicular condition, environmental factors, road geometry, functional properties of the road, etc.). The same methodology can be adopted for developing models for other types of roads (e.g., divided roads, multilane highways, etc.).

Author Contributions

The authors confirm contribution to the paper as follows: study conception and design: Gowri Asaithambi and Vipin; data collection: Vipin; analysis and interpretation of results: Gowri Asaithambi and Vipin; draft manuscript preparation: Vipin and Gowri Asaithambi. All authors reviewed the results and approved the final version of the manuscript.

References

1. *Highway Capacity Manual 2010*. Transportation Research Board of the National Academies, Washington, D.C., 2010.
2. Bang, K. L. *Impact of Side Friction on Speed-Flow Relationships for Rural and Urban Highways*. Technical Report. SWEROAD Indonesia, Bandung, Indonesia, 1995.
3. Dixon, K. K., C. H. Wu, W. Sarasua, and J. Daniel. Posted and Free-flow Speeds for Rural Multilane Highways in Georgia. *Journal of Transportation Engineering*, Vol. 125, No. 6, 1999, pp. 487–494.
4. Ye, Q., A. Tarko, and K. C. Sinha. Model of Free-Flow Speed for Indiana Arterial Roads. *Transportation Research Record: Journal of the Transportation Research Board*, 2001. 1776: 189–193.
5. Kyte, M., Z. Khatib, P. Shannon, and F. Kitchener. Effect of Environmental Factors on Free-Flow Speed. *Transportation Research Circular E-C018*, 2005, pp. 108–119.
6. Hablas, H. E. *A Study of Inclement Weather Impacts on Freeway Free-Flow Speed*. Technical Report. Civil and Environmental Engineering, Virginia Polytechnic Institute and State University, Blacksburg, Virginia, 2007.
7. Wang, L. X. The Study of Establishing Realistic Road Speed Limits. *Research Journal of Applied Sciences Engineering and Technology*, Vol. 4, No. 6, 2012, pp. 599–602.
8. Hashim, I. H. Analysis of Speed Characteristics for Rural Two-lane Roads: A Field Study from Minoufiya Governorate, Egypt. *Ain Shams Engineering Journal*, 2011, pp. 43–52.
9. Azai, T. A., and L. L. Vien. Factors Affecting Free-Flow Speed at Multilane Highways in Malaysia. *Caspian Journal of Applied Sciences Research*, Vol. 2(*AICCE'12 & GIZ' 12*), 2013, pp. 28–35.
10. Moses, R., and E. Mtoi. *Evaluation of Free Flow Speeds on Interrupted Flow Facilities*. Technical Report. Department of Civil Engineering, FAMU-FSU College of Engineering, Tallahassee, Fla., 2013.
11. Wang, T., J. Harvey, J. D. Lea, and C. Kim. *Impact of Pavement Roughness on Vehicle Free-Flow Speed*. Technical Memorandum, California Department of Transportation Division of Research, Innovation and System Information, 2013.
12. Andrade, G. R., C. S. Pitombo, A. L. Cunha, and J. R. Setti. A Model for Estimating Free-Flow Speed on Brazilian Expressways. *Transportation Research Procedia*, Vol. 15, 2016, pp. 378–388.
13. Ma, Y., Y. Zeng, and X. Yang. (2010). Impact of Lane Width on Vehicle Speed of Urban Arterials. In *ICCTP: Integrated Transportation Systems: Green, Intelligent, Reliable*. American Society of Civil Engineers, Beijing, China, 2010, pp. 1844–1852.
14. Tunde, I. Y. The Factors for Free Flow Speed on Urban Arterials – Empirical Evidences from Nigeria. *Journal of American Science*, Vol. 6, No. 12, 2010, pp. 1487–1497.
15. Tseng, P. Y., F. B. Lin, and C. W. Chang. Analysis of Free-flow Speed Characteristics of Urban Arterials. *Asian Transport Studies*, Vol. 2, No. 4, 2013, pp. 363–378.
16. Puan, O. C., and U. T. Abdurrahman. Application of Moving Car Observer Method for Measuring Free Flow Speed on Two-lane Highways. *Jurnal Teknologi*, Vol. 69, No. 2, 2014, pp. 15–19.
17. Balakrishnan, S., and R. Sivanandan. Influence of lane and vehicle subclass on free-flow speeds for urban roads in heterogeneous traffic. *Transportation Research Procedia*, Vol. 10, 2015, pp. 166–175.
18. Rao, A. M., and K. R. Rao. Free Speed Modeling for Urban Arterials - A Case Study on Delhi. *Periodica Polytechnica Transportation Engineering*, Vol. 43, No. 3, 2015, pp. 111–119.
19. Yusuf, I. T., O. O. Adeleke, A. W. Salami, and A. M. Ayanshola. The Factors that Affect the Free Flow Speed on an Arterial in Ilorin, Nigeria. *Nigerian Journal of Technology*, Vol. 35, No. 3, 2016, pp. 473–480.

The Standing Committee on Highway Capacity and Quality of Service (AHB40) peer-reviewed this paper (18-01831).

Article

TRR
JOURNAL OF THE TRANSPORTATION RESEARCH BOARD

Transportation Research Record
2018, Vol. 2672(15) 118–130

DOI: 10.1177/0361198118796716
journals.sagepub.com/home/trr
SAGE

Assessing Highway Travel Time Reliability using Probe Vehicle Data

Piotr Olszewski[1], Tomasz Dybicz[1], Kazimierz Jamroz[2], Wojciech Kustra[2], and Aleksandra Romanowska[2]

Abstract

Probe vehicle data (also known as "floating car data") can be used to analyze travel time reliability of an existing road corridor in order to determine where, when, and how often traffic congestion occurs at particular road segments. The aim of the study is to find the best reliability performance measures for assessing congestion frequency and severity based on probe data. Pilot surveys conducted on A2 motorway in Poland confirm the usefulness and reasonable accuracy of probe data for measuring speed variation in both congested and free-flowing traffic. Historical probe vehicle data and traditional traffic counts from Polish S6 expressway were used to analyze travel time reliability on its 24 road sections. Travel time indexes and reliability ratings for the whole year 2016 were calculated to identify segments with lower reliability and higher expected delay. It is concluded that unlike the HCM-6 method, travel times obtained from probe data should be averaged in 1-hour intervals. Delay index is proposed as a new reliability indicator for road segments. Delay map diagrams are recommended for showing how the congestion spots move in space and with time of day.

Probe vehicle data (PVD, also known as "floating car data"), is a method of measuring traffic speed based on time-space data transmitted from GPS receivers and navigation devices in vehicles on the road. This method of obtaining information about traffic conditions in real time is used by navigation service providers and to aid traffic management. Long-term observations can be utilized to analyze highway travel time reliability; that is, to find how often traffic congestion occurs on a particular road segment.

The paper presents intermediate results of research project MOP-DZ: "Modern methods of calculating road capacity and assessment of traffic conditions on roads outside urban areas and express roads." The aim of the project is to update the Polish methods for evaluating traffic performance and estimating capacity for uninterrupted flow facilities. The project is jointly financed by the General Directorate for National Roads and Motorways and the Polish National Centre for Research and Development. It is being carried out by a consortium of three Universities: Cracow University of Technology, Gdansk University of Technology, and Warsaw University of Technology.

The project uses data from various sources, including traffic volume and speed data from permanent count stations, field surveys, and PVD. This last data category proved to be very useful for illustrating the emergence, spreading, and dissipation of congested regions as well as for assessing travel time reliability. Historical data from 16 weeks in 2015/16 for all national roads and motorways in Poland were used to examine travel time variability and to analyze how often unacceptable traffic conditions occur. The paper presents detailed analysis of a 36.5 km-long segment of S6 expressway, known as the Tricity Bypass Road. The study combined PVD with traffic counts from a permanent count station located at the middle section of the expressway.

Background

From motorists' point of view, variability of travel time is an important measure of service quality of the road network. Travel time distribution can be used to quantify the effect of congestion both recurring daily and non-recurring which is attributable to incidents. Traditionally, level of service measures for multilane roads defined in the *U.S. Highway Capacity Manual*

[1]Civil Engineering, Warsaw University of Technology, Warszawa, Poland
[2]Civil and Environmental Engineering, Gdansk University of Technology, Gdansk, Poland

Corresponding Author:
Address correspondence to Piotr Olszewski: p.olszewski@il.pw.edu.pl

were based on traffic density and did not consider travel time variability and non-recurring congestion. However, the new HCM-6 edition incorporates a new chapter on network reliability (*1*). Similarly, an approach to extend the *German Highway Capacity Manual* (HBS) with highway reliability analysis was proposed by Hartmann et al. (*2*).

Research on the issue of travel time reliability and its effects on driver behavior has been conducted for several years. The results were published as SHRP Report (*3*) and discussed in several papers (*4–7*). Several measures of reliability were defined and proposed for general use. Most of these are based on the concept of Travel Time Index (TTI) which is the ratio of average travel time on a road segment to free-flow travel time. TTI tells us how much longer a trip takes compared with the best conditions. TTI averaged over a time interval (15 min or 1 hour) is a random variable and its distribution can be described statistically. According to HCM (*1*), the most important indicators expressing travel time variability are:

- TTI_{mean}—the mean value of TTI which tells us how much longer, on average, it takes to traverse a road segment, compared with free-flow conditions.
- TTI_x—x-th percentile of the TTI distribution, commonly used x values are: 50%, 80%, and 95%. TTI_{95} is also known as "Planning Time Index," as it is argued that it represents the travel time which should be used for planning a trip if we want to arrive on time with 95% probability.
- Misery index (MI)—is the average TTI for the worst 5% of travel times. It represents the travel time multiplier for the worst-off travelers who are in congestion which happens approximately once a month (1 in 20 chance).
- Reliability rating—is the percentage of vehicle kilometers (or miles) traveled when drivers experience TTI of less than 1.33. The value of 1.33 reflects the subjective threshold beyond which travel time becomes unacceptably prolonged.

An assessment procedure for freeways based on the above measures is proposed in the new HCM-6. However, it is meant not so much for existing roads as for planned facilities, and the distribution of predicted travel times has to be generated based on many assumptions and complex models.

There are a few examples of travel time reliability analyses in the literature based on empirical data (*8*, *9*). These data are collected from permanent traffic count stations which were installed on many major roads as part of intelligent transport systems (ITS) projects. The problem with using count station data is that segment travel times have to be estimated based on vehicle speeds that are measured at one or more spots.

Another approach is to obtain travel times from PVD which are captured over the whole road network and therefore are more representative for road segments. Several examples of application of PVD for forecasting travel time reliability and monitoring traffic conditions can be found in the literature (*9–16*). Most of the studies are applied to uninterrupted flow facilities, and some of them cover urban road networks (*14–17*). In addition to performance measures listed above, other congestion indicators like Average Congestion Index or Congestion Travel Rate were proposed (*17*). Some of the research studies on the use of PVD (*9*, *11*) were concerned with estimation of travel times in real time to be used in driver information systems.

The problem with PVD is that information from only a sample of vehicles is collected, and thus speed data accuracy has to be monitored. In some of the studies (*9*, *11*), results from PVD analyses were compared with loop detector speed data, showing a high match rate. Another problem is that PVD does not include traffic volume data, which has to be obtained using traditional methods or from permanent traffic count stations. Thus, in some studies both data sources are used or combined in order to enhance the accuracy of travel time estimates (*9*). A similar approach has been adopted in the current study.

Although HCM (*1*) defines reliability performance measures for facilities and not for individual road segments, segment data are used throughout the HCM analysis procedure. Analysis of performance of road segments has its merits as it can help to identify capacity deficiency and bottlenecks located downstream of the congested sections.

Measuring Traffic Speed with Probe Vehicle Data

Probe vehicles are equipped with GPS receivers and two-way communication to receive signals from earth-orbiting satellites. The positional information determined from the GPS signals is transmitted to a control center to display the real-time position of the probe vehicles. Travel time information can be determined from the collected data. It is used to obtain information about traffic conditions in real time to aid traffic management. The system has several advantages:

- Relatively low operating cost after initial installation.
- Provides detailed data that are collected continuously over the entire road network.
- Data collection is automated.

The GPS probe vehicle technique has the following disadvantages:

- Privacy issues are a concern for motorists.
- Signals can be lost in urban areas because of buildings, tunnels, or parking garages.
- Relatively high installation cost.

Long-term observations can help to analyze reliability—to find how often traffic congestion occurs on a particular road section. The problem is to find the best method of data collection and analysis for identification of both recurring and non-recurring congestion.

PVD supplier for the MOP-DZ project is CE-Traffic—a Polish "intelligent mobility" company which provides navigation services and ITS traffic management systems. Vehicles equipped with GPS fleet and navigation devices monitor their location, speed, driving direction, and time with a frequency 5–60 s. PVD can be aggregated and compiled in several ways including:

- Origin–Destination traces—cars and goods vehicles,
- Spatiotemporal speed maps (light and heavy vehicles can be compiled separately),
- Speed data in tables or on maps for individual road segments and time intervals, (minimum segment length 100 m, minimum time interval 5–10 min).

Pilot Survey on A2 Motorway Near Warsaw

PVD comprises spatial and temporal data from a sample of vehicles on the road. Accuracy and representativeness of these data decreases as shorter road segments and smaller time intervals are assessed. In order to check the accuracy of PVD, a pilot survey was conducted on a segment of A2 motorway near Warsaw. The study involved manual travel time measurements and was carried out on Friday 29.04.2016—the day before a long weekend.

Five observation points were established on a 6.1 km-long motorway section. During a period of 7 hours, travel times of at least 15 vehicles were measured in every 5 min interval for each of the four road segments between the observation points. Space mean speeds were calculated and compared with speeds obtained from PVD for the same location and time (Figure 1). During the survey period congestion occurred, so the speeds varied between 20 km/h and 130 km/h.

The speeds were averaged (by calculating a harmonic mean) in 5-min, 15-min and 1-h intervals. Root mean square deviation between the two speed measurements was 11.1% for the 5-min intervals, 7.3% for 15-min, and 5.5% for 1-h intervals. The conclusion is that travel times and speeds calculated based on PVD for 1-h intervals have a reasonable accuracy and can be used to assess traffic conditions and time reliability.

Tricity Bypass S6 Expressway Case Study

Description of the Site

Tricity Bypass Road (expressway S6) is the eastern end segment of the Polish National Road No. 6 which runs along the Baltic coast between the cities of Szczecin and the Tricity Metropolitan Area, comprising the cities of Gdansk, Sopot, and Gdynia. The expressway runs North–South along the edge of the metropolitan area and constitutes a key element of the transportation system in the area (Figure 2). Because of growth of the metropolitan area the road, which was built in 1970s, is no longer a typical bypass road—it carries mostly traffic with origins and destinations inside the urban area.

The expressway segment being analyzed is 36.5 km long and includes 11 interchanges. The road is divided and for most of its length has two lanes in each direction with a hard shoulder of up to 1.5 m width. Only one 4-km section has three lanes. The general speed limit is 120 km/h and 100 km/h at the northernmost Gdynia section (Table 1). As the road serves an elongated metropolitan area with a population of 750,000, it carries high traffic volumes. Annual average daily traffic (AADT) surveyed in 2015 was over 77,000 vehicles at the middle section (*18*). In the summer months traffic volume is higher, and the record observed in July 2016 was almost 101,000 veh/day. Thus, the road is operating close to capacity, which is visible especially during peak periods and in summer. This causes an increase in the number of incidents, which further worsens traffic conditions.

Data from Traffic Count Station

A permanent traffic count station is located in the middle of the expressway (Section 5), between Gdansk Osowa and Gdansk Owczarnia interchanges. The station has four inductance loops per lane installed according to TSL and HA specification. Data are collected continuously and comprise for each vehicle: time of passage, spot speed, class, and length.

The data set for the whole year of 2016 was analyzed, including two directions: Southbound and Northbound. Over 27 million vehicles were registered by the station during the year. Data were aggregated in 1-h and 24-h intervals. Average daily volumes were 37,000 vehicles Southbound and almost 38,000 Northbound, with between 6.3% and 8.4% of heavy vehicles. During the summer months traffic was 20% higher than the annual average. Figure 3 presents hourly traffic flow and speed observations for the whole year. Maximum hourly traffic flow was 3,700 veh/h. Average speeds were calculated

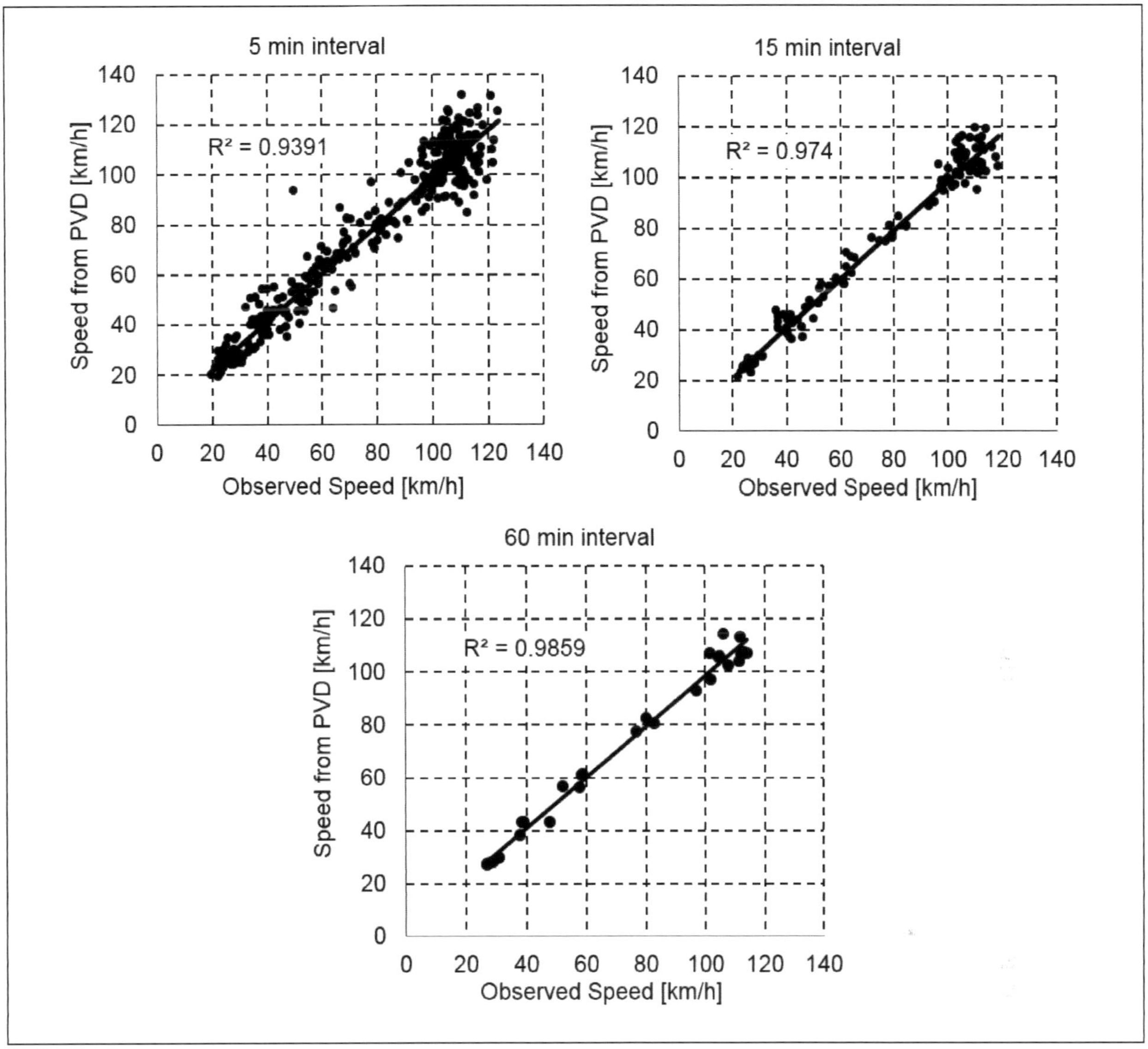

Figure 1. Comparison of PVD and observed speeds.

using the harmonic mean formula. The overall mean speed was 94 km/h Southbound and 92.4 km/h Northbound. Based on vehicle headways (over 6 s in front and over 4 s behind) and maximum hourly flow (below 1,000 veh/h), free-flow speed (FFS) was determined for passenger cars as 113.4 km/h (Southbound) and 111.6 km/h (Northbound). Heavy vehicles were excluded from the calculation as they are subject to a general speed limit of 80 km/h on all divided highways in Poland.

Distribution of traffic according to the time of day on weekdays is shown in Figure 4. There seems to be very little difference between the two directions. Afternoon peak occurs from 1600 to 1700 hours and is slightly higher than the morning peak between 0700 and 0900 hours. On working days traffic flows above 1,000 veh/h occur during 16-h period between 0600 and 2200 hours. This period was selected for the subsequent reliability analysis.

Probe Vehicle Data for S6 Expressway

Probe data were obtained for S6 expressway for the whole year 2016. The 36.5 km-long expressway was divided into 24 sections between interchanges—each direction was a separate section (see Figure 2).

Probe data provided included speeds of light and heavy vehicles separately, aggregated every hour in 2016 for all 24 expressway sections. It included also FFS for

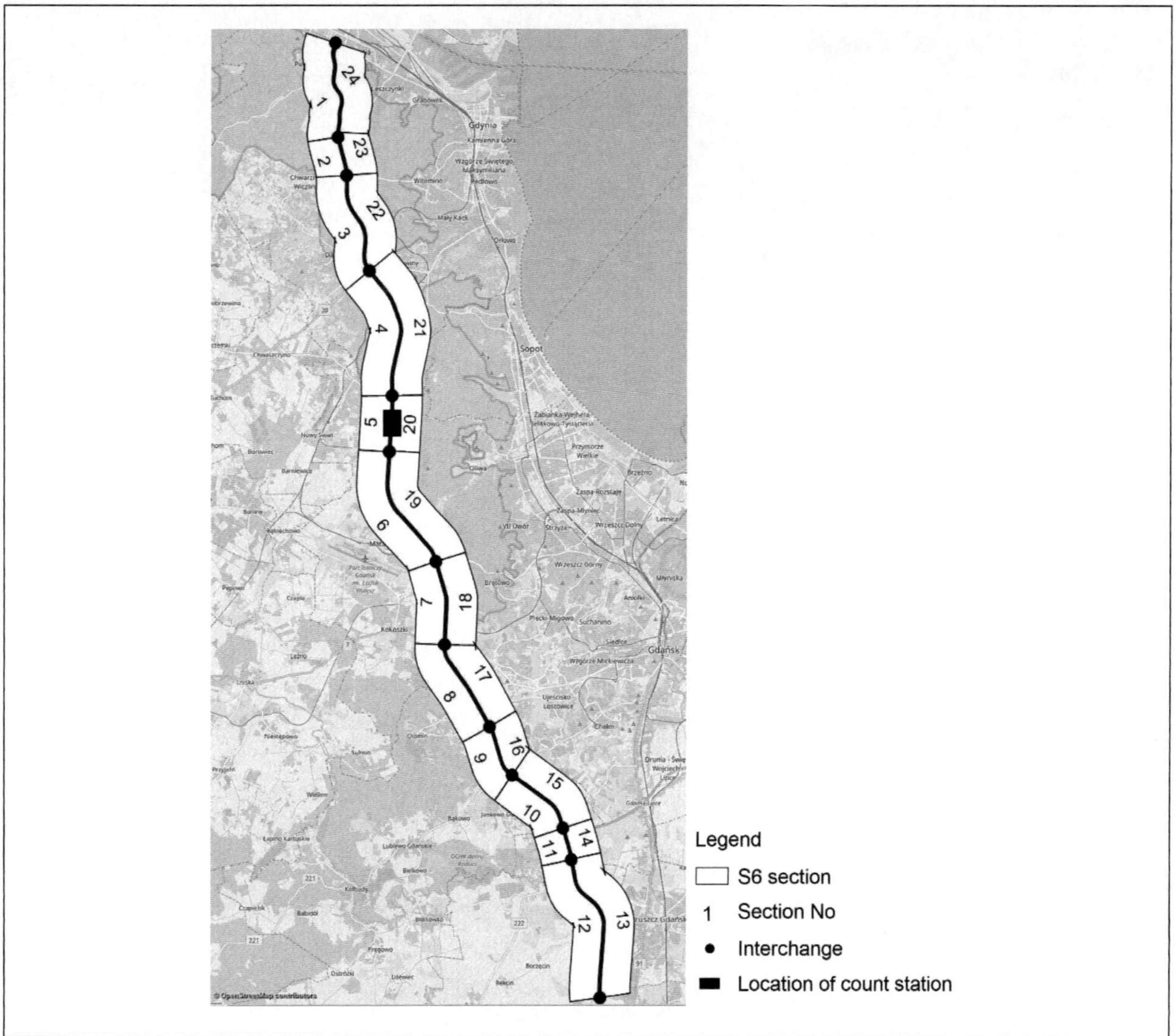

Figure 2. Location of S6 expressway in the Tricity Metropolitan Area.

every section. By combining probe data with traffic volume from the count station, the average traffic speed of all vehicles was calculated for each section and each hour of the year.

Common approach to calculate FFS is to use 85th percentile speed in the off-peak period. The off-peak period can be defined either by time (e.g., overnight or midday) or traffic volume (*19*). However, there is a problem with defining an off-peak period which is to be applied to a large-scale network. First of all, start and end times of peak periods may differ significantly between different locations. Second, a fixed time period does not take into account the fact that adverse weather conditions and incidents can affect traffic flow even in off-peak periods.

Therefore, a new approach to finding off-peak periods and FFS with the use of PVD is proposed. The approach assumes that the off-peak periods can be defined as periods when the average speed is higher than 80% of FFS. Because at first FFS is unknown, it should be approximated by FFS_0, calculated from the whole available PVD speed data set for the analyzed location (without limiting it to off-peak periods). Then the following calculations shall be performed in iterations to limit the data set to the off-peak periods only and to calculate the final FFS value.

Table 1. Traffic Flow and Speed Characteristics of S6 Expressway Sections in 2016

Direction	Section i	Length [km]	AADT [veh/day/direction]	FFS [km/h]	Speed limit, cars [km/h]	Speed limit, HGV [km/h]
Southbound	1	3.50	14,830	104.7	100	80
	2	1.41	32,800	101.3	100	80
	3	3.64	30,779	110.7	100	80
	4	4.85	30,779	112.1	100	80
	5	2.01	36,969	112.1	120	80
	6	4.60	36,969	113.5	120	80
	7	3.03	38,569	111.7	120	80
	8	3.43	34,482	115.2	120	80
	9	1.99	34,482	113.1	120	80
	10	2.78	34,482	113.0	120	80
	11	1.18	18,934	121.8	120	80
	12	5.40	17,842	119.8	120	80
Northbound	13	5.41	17,842	118.3	120	80
	14	1.16	18,934	118.0	120	80
	15	2.79	34,482	109.0	120	80
	16	1.99	34,482	109.6	120	80
	17	3.43	34,482	109.8	120	80
	18	3.03	38,569	109.5	120	80
	19	4.64	36,969	114.4	120	80
	20	1.97	36,969	112.0	120	80
	21	4.85	30,779	115.4	100	80
	22	3.64	30,779	112.2	100	80
	23	1.58	32,800	112.1	100	80
	24	3.30	14,830	105.8	100	80

Note: Traffic data including traffic volume were obtained from National Census Traffic Survey and count station owned by the road authority (*18*). Speed data including FFS were obtained from CE-Traffic, PVD data supplier.

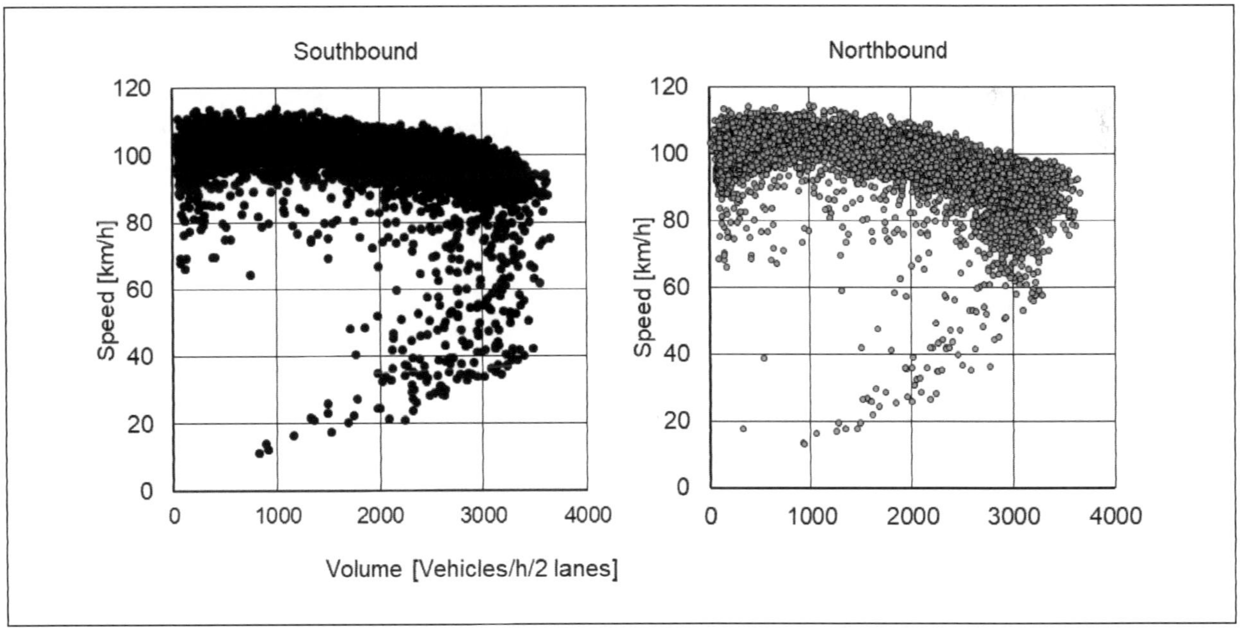

Figure 3. Hourly traffic flow and speed observations in the year 2016.

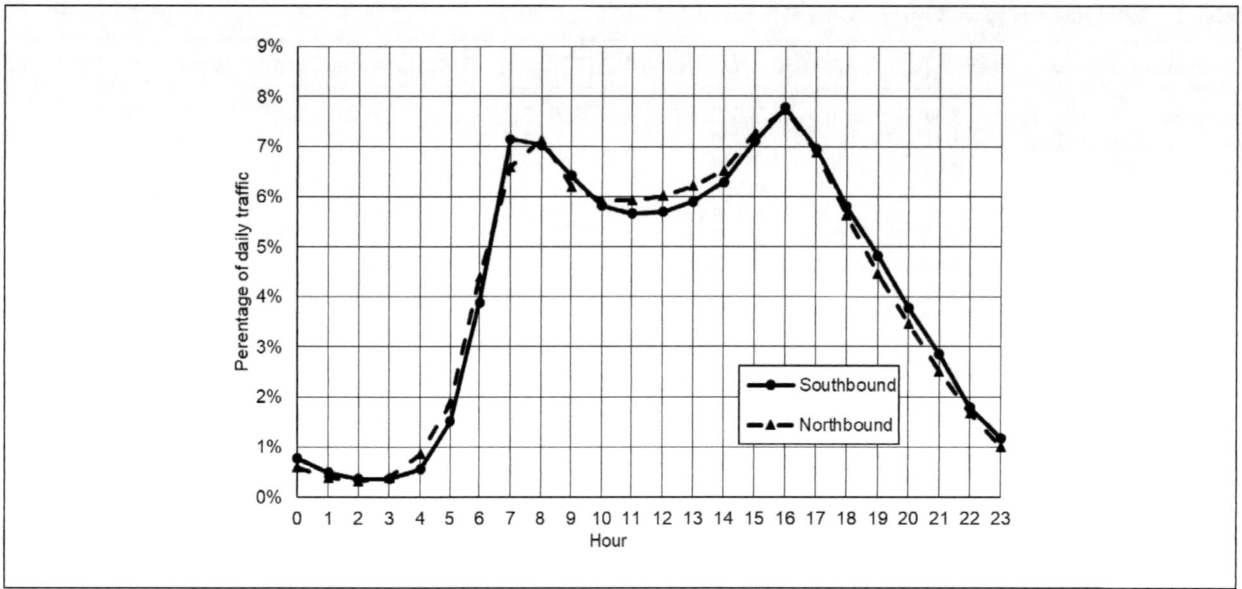

Figure 4. Hourly distribution of traffic flow at the count station on a weekday.

- Threshold speed value St_i equal to 80% of FFS_{i-1} is calculated.
- Time periods with average speed $< St_i$ are removed from the dataset (assumed not to be off-peak or uncongested periods).
- FFS_i is calculated as 85th percentile from the new data set.

The proposed method was chosen because it allows full utilization of available PVD dataset without limiting it to only short, arbitrarily defined fixed time periods. Moreover, the method is self-contained in a sense that it is based solely on the speed data and does not require either traffic volume information or any location-specific attributes. Therefore, it seems to be a well-suited choice for calculation of FFS based on PVD for a large-scale network. Parameters of the method can be adjusted and calibrated for a particular PVD data source.

Travel Time Reliability Analysis

The next step was to conduct reliability analysis for the whole year 2016, based on section travel times obtained from probe data. The reliability reporting period (RRP) included 252 working weekdays in 2016. Weekends and public holidays were excluded. In addition, the analysis was limited to a 16-h period (0600–2200 hours) per day, excluding the nighttime low-traffic period (see Figure 4).

TTI values were calculated for all the 4,032 hours (252 days*16) and 24 expressway sections. Figure 5 shows the distributions of TTI values for selected sections. The distributions have long tails, which indicate that there are infrequent periods of congestion when travel times are more than twice the free-flow time.

Table 2 shows the parameters of TTI distribution for all the sections. Generally, the assessment is rather good. For all the expressway sections both TTI_{mean} and TTI_{80} values are below the 1.33 threshold. However, most of the TTI_{95} values are above 1.33. Differences between segments are visible in TTI_{95} and in the MI (Figure 6). The highest MI value of 4.08 occurs at section 15, which is the southern approach to Gdansk. Based on the MI, the Northbound direction performs worse than the Southbound: there are five sections with MI >3.0, compared with only one in the opposite direction.

Reliability rating values shown in Table 2 were calculated in two ways. Vehicle-kilometers traveled (VKT)-based reliability is expressed by the formula:

$$RR1_i = \frac{\sum_{p \in P1.33} VKT_{ip}}{\sum_p VKT_{ip}} 100\% \quad p \in RRP \tag{1}$$

where:

VKT_{ip} = vehicle kilometers traveled on section "i" during analysis period "p"

p = analysis period: 1 h belonging to the RRP

P1.33 = set of all analysis periods when TTI < 1.33

Vehicle kilometers traveled on section "i" during hour "p" are equal to:

$$VKT_{ip} = Vol_{sp} \frac{AADT_i}{AADT_s} L_i \tag{2}$$

where:

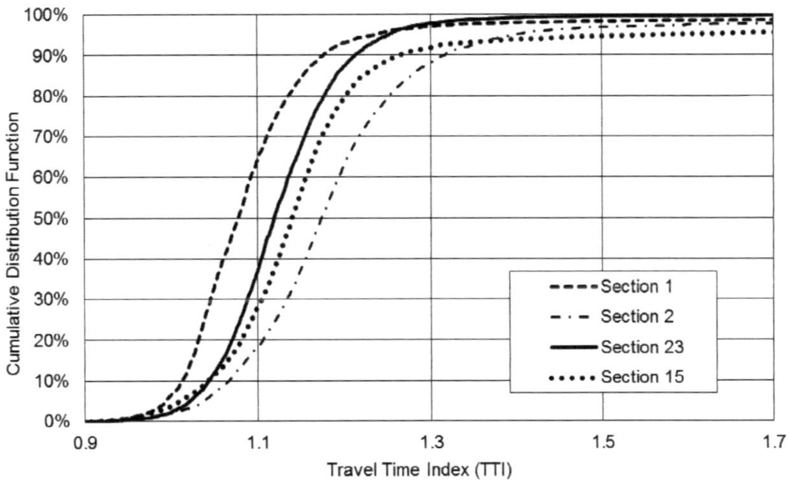

Figure 5. Cumulative distributions of travel time index for selected sections.

Table 2. Reliability Analysis with PVD Speed Data Application

Direction	Section i	TTI_{mean}	TTI_{80}	TTI_{95}	Misery index	Reliability rating by veh-km	Reliability rating by time
Southbound	1	1.14	1.14	1.24	2.37	97%	97%
	2	1.24	1.25	1.40	2.49	89%	91%
	3	1.19	1.17	1.28	2.74	96%	96%
	4	1.18	1.17	1.29	2.50	96%	96%
	5	1.23	1.20	1.43	3.10	92%	93%
	6	1.19	1.18	1.46	2.34	92%	93%
	7	1.13	1.15	1.25	1.70	97%	97%
	8	1.21	1.17	1.39	2.97	93%	94%
	9	1.15	1.18	1.29	1.74	96%	96%
	10	1.14	1.16	1.24	1.61	98%	98%
	11	1.17	1.18	1.39	2.00	93%	94%
	12	1.17	1.20	1.36	1.80	93%	94%
Northbound	13	1.14	1.18	1.25	1.61	98%	98%
	14	1.20	1.18	1.28	2.88	97%	97%
	15	1.28	1.20	1.60	4.08	92%	93%
	16	1.26	1.19	1.48	3.83	92%	93%
	17	1.18	1.21	1.35	1.98	93%	94%
	18	1.23	1.17	1.37	3.51	93%	94%
	19	1.27	1.22	1.88	3.26	86%	88%
	20	1.23	1.22	1.41	2.80	91%	92%
	21	1.22	1.20	1.38	2.97	93%	94%
	22	1.16	1.16	1.27	2.10	96%	97%
	23	1.13	1.18	1.25	1.49	98%	98%
	24	1.19	1.13	1.43	3.31	93%	94%

Vol_{sp} = hourly traffic volume measured at the count station during period p,

$AADT_i$ = annual average daily traffic for section "i"

$AADT_s$ = annual average daily traffic for section "s" (count station)

L_i = length of section "i"

Time-based reliability rating is simply the percentage of analysis periods when TTI <1.33:

$$RR2_i = \frac{N_{ip \in P1.33}}{N_{ip}} 100\% \qquad (3)$$

where:

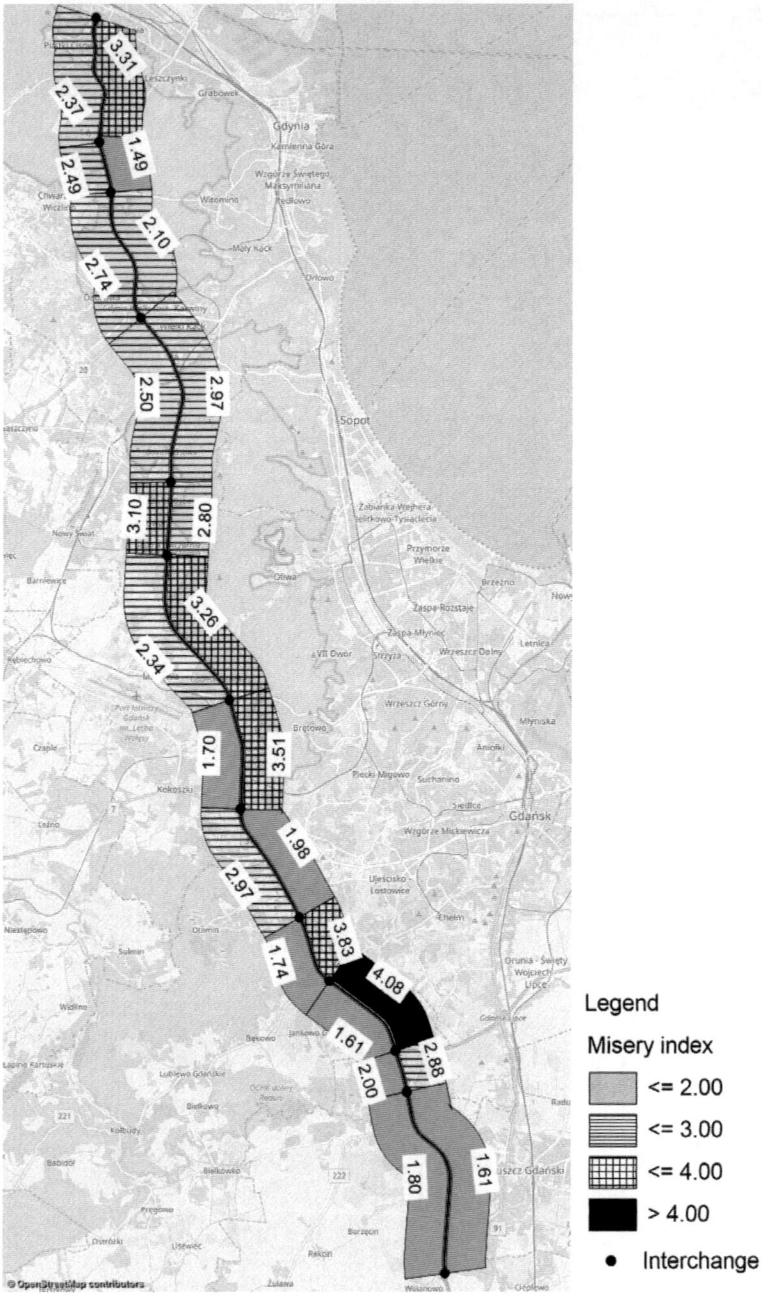

Figure 6. Misery index values for S6 expressway sections.

N_{ip} = total number of analysis periods for section "i"
$N_{ip \in P1.33}$ = number of analysis periods for section "i" when TTI < 1.33

Table 2 shows that both reliability rating formulations result in values which are only marginally different. Therefore, time-based reliability rating (Equation 3) can be used when traffic volume data are not available. However, it should be noted that VKT-based formula is more accurate and should be used whenever possible, especially in situations when traffic volumes are highly variable.

Generally speaking, the reliability rating of the expressway is very good. Only two sections, 2 and 19,

Direction	Northbound – Section No.											
Hour	13	14	15	16	17	18	19	20	21	22	23	24
0:00	15%	14%	6%	7%	8%	13%	6%	6%	13%	8%	10%	5%
1:00	17%	15%	8%	9%	11%	11%	10%	10%	17%	13%	13%	5%
2:00	19%	19%	12%	14%	16%	13%	12%	11%	17%	14%	14%	6%
3:00	25%	24%	16%	15%	17%	13%	13%	12%	18%	14%	16%	10%
4:00	26%	27%	18%	16%	20%	21%	20%	20%	22%	16%	12%	6%
5:00	30%	30%	18%	15%	19%	16%	12%	12%	18%	17%	17%	10%
6:00	21%	24%	16%	14%	18%	15%	13%	14%	18%	14%	18%	9%
7:00	18%	21%	18%	15%	15%	11%	13%	15%	18%	15%	17%	9%
8:00	16%	17%	36%	25%	19%	11%	18%	21%	23%	14%	15%	13%
9:00	16%	14%	32%	35%	24%	15%	20%	23%	28%	17%	14%	31%
10:00	19%	32%	79%	78%	37%	12%	26%	26%	48%	41%	16%	37%
11:00	40%	190%	131%	102%	39%	14%	38%	35%	30%	16%	15%	98%
12:00	22%	111%	123%	110%	53%	14%	30%	36%	23%	13%	16%	136%
13:00	14%	50%	84%	82%	37%	31%	59%	45%	21%	14%	14%	26%
14:00	13%	54%	51%	35%	23%	31%	37%	36%	26%	13%	13%	23%
15:00	12%	21%	22%	17%	18%	12%	30%	23%	26%	14%	21%	37%
16:00	12%	11%	30%	26%	21%	20%	44%	51%	55%	14%	13%	54%
17:00	15%	10%	62%	31%	21%	141%	124%	75%	54%	14%	11%	88%
18:00	14%	8%	41%	34%	22%	181%	75%	47%	41%	12%	12%	93%
19:00	8%	6%	10%	15%	10%	87%	44%	30%	30%	10%	11%	79%
20:00	7%	6%	9%	8%	10%	6%	15%	14%	18%	8%	10%	51%
21:00	7%	7%	7%	6%	7%	6%	9%	11%	30%	7%	8%	4%
22:00	11%	27%	15%	8%	10%	10%	8%	10%	43%	9%	9%	10%
23:00	11%	47%	11%	6%	8%	7%	8%	8%	16%	9%	11%	12%

Figure 7. Delay map for S6 Northbound—weekdays in August 2016.

have a rating below 90%. Ranking of sections according to reliability rating is different than that using TTI_{95} and MI. This shows that the measures are complementary in identifying problem areas.

Reliability measures in Table 2 indicate the location of congested areas but are not time-specific. To show how congestion varies over the time of day we need to look at how travel times change both in time and space. A new indicator called "Delay Index" (DI) is proposed, defined as follows:

$$DI_{ij} = \left(TTI_{\text{mean}\,i,j} - 1\right)*100\% \qquad (4)$$

where:

$TTI_{\text{mean}\ i,j}$ = mean value of TTI for section i and hour j

The DI tells us how much extra time is needed to traverse a road section at a particular time of day, compared with the free-flow travel time. Simply put, it is the average delay expressed as a percentage of the free-flow travel time.

Delay indices can be displayed in the form of "Delay maps"—showing how DI values along the road corridor change with time. Delay maps for S6 expressway are shown in Figures 7 (Northbound) and 8 (Southbound) for the month of August 2016. The summer months are special because tourist traffic to and from the Baltic coast mixes with the regular users of the expressway. Figure 7 shows that in the Northbound direction there is no morning peak—congestion starts to build up around 1100 hours and later in the evening after 1700 hours. At

Direction	Southbound - Section No.											
Hour	1	2	3	4	5	6	7	8	9	10	11	12
0:00	12%	13%	15%	15%	12%	13%	11%	14%	11%	12%	19%	23%
1:00	8%	14%	13%	13%	14%	11%	14%	15%	14%	13%	16%	21%
2:00	15%	17%	18%	20%	17%	17%	16%	14%	12%	11%	14%	18%
3:00	12%	13%	11%	11%	8%	10%	10%	13%	13%	9%	12%	15%
4:00	19%	17%	14%	13%	10%	12%	6%	17%	13%	13%	10%	14%
5:00	23%	13%	13%	14%	10%	12%	8%	8%	8%	7%	9%	10%
6:00	7%	5%	5%	8%	5%	6%	4%	5%	4%	5%	6%	5%
7:00	5%	9%	6%	11%	10%	9%	3%	4%	5%	5%	6%	7%
8:00	8%	15%	9%	14%	11%	12%	6%	7%	8%	10%	7%	16%
9:00	5%	17%	10%	15%	16%	13%	6%	9%	10%	11%	10%	12%
10:00	10%	20%	13%	16%	16%	14%	8%	10%	12%	12%	12%	13%
11:00	12%	21%	14%	19%	23%	18%	12%	14%	14%	17%	13%	13%
12:00	13%	27%	16%	22%	36%	23%	12%	15%	16%	16%	14%	21%
13:00	12%	27%	15%	32%	60%	32%	12%	16%	17%	15%	16%	40%
14:00	12%	29%	17%	32%	52%	31%	18%	56%	38%	15%	14%	31%
15:00	17%	35%	19%	42%	55%	35%	16%	60%	28%	19%	13%	21%
16:00	49%	75%	137%	84%	34%	70%	19%	69%	23%	15%	17%	34%
17:00	153%	80%	86%	61%	90%	65%	19%	71%	22%	15%	32%	36%
18:00	74%	76%	97%	66%	80%	63%	14%	64%	19%	15%	18%	22%
19:00	23%	32%	42%	37%	47%	34%	12%	64%	23%	14%	12%	18%
20:00	10%	11%	12%	22%	25%	17%	10%	38%	33%	11%	11%	11%
21:00	8%	11%	9%	11%	9%	9%	10%	15%	13%	11%	12%	12%
22:00	13%	16%	13%	17%	13%	13%	13%	15%	13%	14%	17%	18%
23:00	8%	9%	10%	13%	11%	11%	9%	12%	11%	10%	14%	16%

Figure 8. Delay map for S6 Southbound—weekdays in August 2016.

these hours on some sections the average travel time is between two and three times the free-flow time (i.e., DI is between 100% and 200%). In the Southbound direction the most congested period is after 1600 when tourists start returning from the costal resorts. It extends for more than half of the expressway length.

Discussion

One of the issues that needs discussion is estimating the average FFS. An individual free-flowing vehicle is a car which travels without interaction with other cars. Such a situation can be defined in terms of headways: the condition for a free movement is a headway of 5–7 s at the front and a headway of 4–6 s behind the vehicle. With such assumptions, the traffic volume should not exceed 300–400 veh/h. Different procedures and models for estimating the mean free velocities (*19*, *20*) have been developed. Pilot studies conducted in Poland show that the following conditions affect the average FFS: road type, posted speed limit, type of vehicle, time of day (day, night), and geometric parameters of the road. These issues will be addressed in future research.

Determination of FFS from PVD is another problem. In the absence of flow data, it can only be based on the distribution of observed speeds or mean speeds for the selected aggregation period (e.g., 1 h). An important question is: what to do when the FFS determined from empirical data is higher than the posted speed limit? Which value should be taken as the basis for TTI calculation? It can be argued that assessment should reflect reality and therefore empirically determined FFS is a sound reference basis for TTI.

Although the use of PVD for travel time reliability assessment is relatively easy and cost efficient, the method faces several problems. Firstly, as PVD are based on a sample of vehicle probes, accuracy of speed data from PVD is not guaranteed and needs to be monitored. When traffic volume is low, there may be time periods with insufficient probes to estimate a reliable value of speed. So far, the Polish experience suggests using hourly intervals for speed and travel time averaging. In the future, as market penetration of GPS equipment increases, it may be possible to go down to 15-min averaging.

When conducting a travel time reliability study, selection of the RRP is an important decision. It would seem that one full year would be the best choice. However, traffic conditions vary considerably with seasons and also with days of the week. Therefore, it would be best to assess separately highway reliability during working days and during weekends and public holidays. Periods of special interest such as summer holidays can also be evaluated separately.

Which reliability measures are the most appropriate for use? These may vary according to the geographical and time scope of the assessment. Measures recommended in the new HCM-6 such as TTI percentiles, MI and reliability rating should be the basis for any reliability assessment. DIs may be more appropriate for showing variability in both time and space.

Conclusions

Results of pilot surveys conducted on A2 motorway and S6 expressway confirm the usefulness of PVD for assessing existing traffic conditions and identifying the location of congested regions. At this time, market penetration rate of GPS equipment allows for probe data to be used to obtain reliable traffic speed values averaged in 1-h intervals. Historical PVD data sets can then be used to analyze how often and where unacceptable traffic conditions occur in real life. With proper data filtering and processing, the method helps to assess the actual travel time reliability of a road network and to identify segments with capacity deficiency.

The proposed "Delay Index" is a useful indicator of how much the actual travel time will be longer than the free-flow time. Delay maps are proposed as a novel way of presenting, in one diagram, where and when to expect delay, and also of what magnitude it is likely to be. They should be of interest both for traffic professionals and for the general public.

Based on a 1-year reporting period considering only weekdays, it can be concluded that the Tricity Bypass Road has good travel time reliability. Out of 24 sections analyzed only two had the reliability rating slightly under 90%, and only one section had the MI over 3. However, during the summer months of July and August, traffic is considerably heavier, as the road is used by many tourists going to holiday resorts located on the Baltic Sea coast. At certain times of the day and in some locations, average travel time is almost tripled compared with the free-flow conditions. It seems that introduction of an ITS active traffic management system is highly recommended and would be of great benefit to the users.

Acknowledgments

This study is part of RID-I-50 OT2-2B/PK-PW-PG research project, financed by the National Centre for Research and Development (Poland) and the General Directorate for National Roads and Motorways. The authors would like to thank all the project partners who contributed to this research. Special thanks are due to CE-Traffic for providing probe vehicle data and helping with their processing.

Author Contributions

The authors confirm contribution to the paper as follows: study conception and design: P. Olszewski, T. Dybicz; data collection: T. Dybicz, W. Kustra, A. Romanowska; analysis and interpretation of results: P. Olszewski, T. Dybicz, W. Kustra, K. Jamroz; draft manuscript preparation: P. Olszewski, A. Romanowska. All authors reviewed the results and approved the final version of the manuscript.

References

1. *Highway Capacity Manual: A Guide for Multimodal Mobility Analysis*, 6th ed. Transportation Research Board, Washington, D.C., 2016.
2. Hartmann, M., P. Vortisch, and B. J. Schroeder. A German Approach to Freeway Facility Evaluation. *Transportation Research Record: Journal of the Transportation Research Board*, 2015. 2483: 66–73.
3. Zegeer, J., J. Bonneson, R. Dowling, P. Ryus, M. Vandehey, W. Kittelson, N. Rouphail, B. Schroeder, A. Hajbabaie, B. Aghdashi, T. Chase, S. Sajjadi, R. Margiotta, and L. Elefteriadou. *SHRP2 Report S2-L08-RW-1: Incorporating Travel Time Reliability into the HCM*. Transportation Research Board of the National Academies, Washington, D.C., 2013.
4. Hamedi, M., A. Haghani, and Y. Zhang. Travel Time Reliability Measurement Sensitivity to Data Source Collection. *Smart Highways*, Vol. 11, 2016.
5. Kaparias, I., M. G. H. Bell, and H. Belzner. A New Measure of Travel Time Reliability for In-Vehicle Navigation Systems. *Journal of Intelligent Transportation Systems*, Vol. 12, No. 4, 2008, p. 10.
6. Lomax, T., D. Schrank, S. Turner, and R. Margiotta. *Selecting Travel Reliability Measures*. Texas Transportation Institute Monograph, 2003.
7. Lyman, K., and R. Bertini. Using Travel Time Reliability Measures to Improve Regional Transportation Planning

and Operations. *Transportation Research Record: Journal of the Transportation Research Board*, 2008. 2046: 1–10.

8. Chen, C., A. Skabardonis, and P. Varaiya. Travel Time Reliability as a Measure of Service. *Transportation Research Record: Journal of the Transportation Research Board*, 2003. 1855: 74–79.
9. Westerman, M., R. Litjens, and J. Linnartz. *Integration of Probe Vehicle and Induction Loop Data: Estimation of Travel Times and Automatic Incident Detection*. California Partners for Advanced Transportation Technology, UC Berkeley, 1997.
10. Wang, Z., A. Goodchild, and E. McCormack. A Methodology for Forecasting Freeway Travel Time Reliability Using GPS Data. *Transportation Research Procedia*, Vol. 25, 2017, pp. 842–852.
11. Wei, W., G. Xiucheng, J. Jing, and R. Bin. GPS Probe Based Freeway Real-Time Travel Speed Estimation Using Kalman Filter. *Proceedings - 2010 International Conference on Intelligent System Design and Engineering Application, ISDEA 2010*, Vol. 1, 2011, pp. 797–800.
12. Pei, X., S. C. Wong, Y. C. Li, and N. N. Sze. Full Bayesian Method for the Development of Speed Models: Applications of GPS Probe Data. *Journal of Transportation Engineering*, Vol. 138, No. 10, 2012, pp. 1188–1195.
13. Izadpanah, P. *Freeway Travel Time Prediction Using Data from Mobile Probes*. University of Waterloo, 2010.
14. Jenelius, E., and H. N. Koutsopoulos. Travel Time Estimation for Urban Road Networks Using Low Frequency Probe Vehicle Data. *Transportation Research Part B: Methodological*, Vol. 53, 2013, pp. 64–81.
15. Zheng, F., and H. Van Zuylen. Urban Link Travel Time Estimation Based on Sparse Probe Vehicle Data. *Transportation Research Part C: Emerging Technologies*, Vol. 31, No. 111, 2013, pp. 145–157.
16. Zhang, K., D. (Jian) Sun, S. Shen, and Y. Zhu. Analyzing Spatiotemporal Congestion Pattern on Urban Roads Based on Taxi GPS Data. *Journal of Transport and Land Use*, Vol. 10, No. 1, 2017, pp. 675–694.
17. Sun, D., X. Liu, A. Ni, and C. Peng. Traffic Congestion Evaluation Method for Urban Arterials: A Case Study of Changzhou, China. *Transportation Research Record: Journal of the Transportation Research Board*, 2014. 2461: 9–15.
18. General Directorate for National Roads and Motorways. *National Roads – General Traffic Survey 2015*. General Directorate for National Roads and Motorways, Warsaw, 2016.
19. Texas Transportation Institute. *The Keys to Estimating Mobility in Urban Areas. A White Paper Prepared for the Urban Transportation Performance Measure Study*. Texas Transportation Institute, 2005.
20. Gaca, S., and M. Kiec. Speed Management for Local and Regional Rural Roads. *Transportation Research Procedia*, Vol. 14, 2016, pp. 4170–4179.

The Standing Committee on Highway Capacity and Quality of Service (AHB40) peer-reviewed this paper (18-03962).

Article

TRR
JOURNAL OF THE TRANSPORTATION RESEARCH BOARD

Transportation Research Record
2018, Vol. 2672(15) 131–141

DOI: 10.1177/0361198118784401
journals.sagepub.com/home/trr
SAGE

Defining Freeway Design Capacity Based on Stochastic Observations

Siavash Shojaat[1], Justin Geistefeldt[2], Scott A. Parr[3], Luis Escobar[4], and Brian Wolshon[1]

Abstract

The estimation of capacity as a parameter to assess traffic flow performance on freeway facilities has received considerable attention in the literature. Despite the general acceptance of the stochastic notion of capacity, limited research has been conducted on how to select a single representative design value from a capacity distribution function. This paper reports the results of an empirical comparison between conventional capacity estimates and those obtained by maximizing the sustained flow index (SFI) for 19 U.S. freeway sections. The SFI is defined as the product of the traffic volume and the probability of survival at this volume. The capacity of each cross-section was estimated by analyzing the speed–flow relationship and applying methods for stochastic capacity analysis. The results show that the optimum volumes obtained by maximizing the SFI estimated in 5-minute intervals correspond well to the 15% probability of breakdown proposed in the *Highway Capacity Manual 6th Edition* to estimate the capacity from field data. From these results, it was concluded that maximizing the SFI can be considered a preferred approach to estimate a single, representative value of freeway capacity.

Capacity is among the most influential parameters used to assess traffic flow performance on freeway facilities. In general, capacity is defined as the maximum volume that can traverse a homogenous freeway section under prevailing traffic and control conditions. In conventional analyses, the capacity of a freeway is treated as a constant value. Thus, operational capacity in this traditional sense can be empirically estimated by identifying the apex volume from a calibrated speed–flow–density model. However, this approach disregards the stochastic nature of capacity. In fact, the apex volume is just one possible estimate for the capacity, because the capacity of a freeway segment can change momentarily as individual driving behaviors also change. If used for capacity analysis, the conventional approach may also be criticized as it does not take into account the breakdown phenomena. Fundamentally, uncongested and congested traffic states represent different capacities as a result of the capacity drop. Fitting the fundamental diagram through both uncongested and congested observations can bias the capacity estimation results.

In contrast to the conventional perception of capacity, many researchers have reported that freeway facilities reach their capacities at different traffic volumes even under similar prevailing conditions (*1–5*). This suggests that capacity is not a fixed value but a random variable from a unique distribution function, subject to the behavioral variability of individual drivers (*6*). Thus, a capacity distribution function can be estimated by examining traffic flow breakdowns. However, it is important to note that freeways are not designed based on all of the observed capacity values. Rather, a single, representative capacity value is typically used for analysis and freeway design in practice.

Despite the wide acceptance of the stochastic nature of capacity and different methods developed to estimate its distribution function, there have been limited efforts to develop an analytical method to compute a single capacity value from the capacity distribution function. To select a single capacity value, some researchers have superimposed the conventional speed–flow diagrams over the capacity distribution functions. From these plots, average probabilities of breakdown corresponding

[1]Gulf Coast Center for Evacuation and Transportation Resiliency, Department of Civil and Environmental Engineering, Louisiana State University, Baton Rouge, LA
[2]Institute for Traffic Engineering and Management, Ruhr-University Bochum, Bochum, Germany
[3]Department of Civil Engineering, Embry Riddle Aeronautical University, Daytona Beach, FL
[4]Department of Experimental Statistics, Louisiana State University, Baton Rouge, LA

Corresponding Author:
Address correspondence to Siavash Shojaat: sshoja1@lsu.edu

to apex volumes are commonly selected as benchmarks (*7*, *8*). However, as a capacity estimated using this approach ultimately relies on a calibrated speed–flow model, it is subject to the same limitations of the conventional approach discussed above.

A recent study by Shojaat et al. introduced the sustained flow index (SFI) as a measure of freeway flow performance (*9*). This measure is defined as the product of the traffic volume and the probability of "survival," that is, the probability that traffic flow does not break down at this volume. A first application based on German freeway data indicated that the optimum volume calculated by maximizing the SFI may be a reasonable freeway design volume. The strength of this approach is that it solely relies on the capacity distribution function representing the risk of a flow breakdown in fluid traffic conditions.

The goal of this paper is to investigate the validity of the optimum volume suggested by Shojaat et al. (*9*) as reasonable capacity estimate. A thorough comparison between optimum volumes and conventional capacity estimates is conducted based on a large sample of U.S. freeway data. The conventional (deterministic) capacity estimates, capacity distribution functions and optimum volumes are estimated and compared for 19 U.S. freeway bottleneck sections. In addition, a new procedure is developed to estimate the Wald and Bootstrap confidence intervals for the optimum volumes and determine their stability. The paper also proposes new mathematical derivations which suggest the probability of breakdown at the optimum volume may solely depend on the shape parameter of the capacity distribution function. This finding indicates that probability of breakdown at the capacity may remain constant for roads with different numbers of lanes. A comparison of these methods shows that the optimum volumes obtained by maximizing the SFI are, on average, 1.2% different from the conventional capacity estimates. It is also observed that the optimum volumes are stable, as their confidence intervals are fairly small. These findings also provide a practical method for estimating the capacity distribution function for sections for which a conventional capacity estimate is already available or for circumstances in which a reliable capacity distribution function cannot be estimated.

The paper starts with a literature review to describe the evolution of capacity estimation methods that have led to this research. Next, a description of the conventional and stochastic methodologies used to estimate capacity values in this research is provided and the procedure to derive the confidence intervals for the optimum volumes is introduced. Then, to demonstrate an application of the experimental method, the proposed methodologies are applied to the selected freeway bottlenecks to investigate similarities between the calculated results. The paper concludes with a summary of the findings and recommendations for future application of this work.

Literature Review

Capacity is traditionally viewed as the maximum volume that can be sustained by the facility. To estimate the capacity of basic freeway segments, the *Highway Capacity Manual* (*HCM*) delivers a set of base capacities dependent on the free-flow speed (*10*). Although these capacity values are found to be fairly representative for U.S. freeways, local conditions may significantly affect the accuracy of the capacity estimates. Thus, many researchers have suggested different mathematical functions to locally determine the relationship between speed, volume, and density. Given the empirical observations of speed, flow, and density, parameters of the predetermined model are calibrated and the apex volume of the speed–flow diagram is considered as the capacity of the segment. Therefore, selecting an appropriate model plays a key role to estimate realistic capacity values.

Van Aerde proposed a mathematical function based on a simple car-following model which assumes density to be a function of the current speed and the free-flow speed (*11*). Although the suggested model is continuous, the assumption of a linear relationship between speed and density is relaxed and, as a result, it is well capable of describing different traffic states. Van Aerde and Rakha suggested a multivariate calibration of the speed–density scatterplot as an unbiased method when it is not clear which one (of the speed, flow, or density) is the dependent and which is the independent variable (*12*).

As the apex volume of the speed–flow diagram is only one possible capacity estimate and freeway facilities may experience saturation at higher or lower volumes, other procedures were developed to estimate the probability of breakdown occurrence as a function of the traffic volume. Van Toorenburg estimated the capacity distribution function based on the analogy drawn between incomplete lifetime data and freeway capacity (*13*). In order to derive the capacity distribution function, observations of both congested and non-congested regimes were included. Brilon et al. used a slightly modified approach and considered only the non-congested observations for the analysis because observations under congested traffic conditions provide no information about the capacity before a breakdown (*14*, *15*). The researchers also employed different parametric distributions to fit the empirical observations collected from German freeways and found that Weibull distribution provides the best fit to the data.

Geistefeldt applied the method implemented by Brilon et al. to estimate the capacity distribution function and superimposed the Van Aerde speed–flow diagrams over the estimated distribution function for 27 German freeway sections (7). It was concluded that for 5-minute aggregation intervals, the average probability of breakdown corresponding to the apex volume of the Van Aerde model is nearly 3%. Modi et al. performed the same analysis and concluded that the peak volume corresponds to nearly 4% breakdown probability for U.S. freeways (8). It was also found that for the sections under investigation the capacities provided by the *HCM* were generally greater than those estimated with other methods.

Shojaat et al. (9) introduced the SFI as a stochastic performance measure obtained by multiplying the traffic volume and the probability of survival, which is the complement of the breakdown probability. The volume at which the maximum SFI is reached was referred to as the optimum volume. This optimum volume is the one that provides the best compromise between the probability of breakdown and the unutilized capacity of freeway. As a result, assuming different capacity distribution function types, the breakdown probability corresponding to the optimum volume that maximizes the SFI was introduced as the benchmark to define capacity.

The literature review suggests that, in spite of the progress made in the field of capacity analysis, few efforts beyond the research conducted by Shojaat et al. have been made to propose an analytical approach to select a single design value from the capacity distribution function. Thus, this research aims to (1) empirically compare the optimum volumes with capacities obtained in the speed–flow diagram by applying the Van Aerde model to investigate the reasonableness of the estimation results, (2) provide confidence intervals for the parameters of the Weibull distribution as well as the optimum volumes to investigate their variability, and (3) develop a method to transform a conventional capacity estimate into an entire capacity distribution function.

Methodology

The research methodology is presented in two parts. The first part discusses the Van Aerde model as a conventional approach to estimate capacity as well as the stochastic approach to estimate a capacity distribution function based on models for censored data. The second part provides a description of the SFI as a method to select a single capacity value from the capacity distribution function. This part also provides an explanation of the methods used to determine confidence intervals for the model parameters and the optimum volumes. Subsequently, the results of the application of the presented methodology to data collected from the 19 freeway bottleneck sections are discussed.

Deterministic Capacity Estimation

To estimate conventional (deterministic) capacities, the Van Aerde model (11), as a function capable of describing the speed–flow–density relationship based on a simple car-following equation, was applied in this study. In this model (Equation 1), the distance headway between consecutive vehicles (h) only depends on the free-flow speed (s_f), the current speed (s), and three parameters (c_1, c_2, c_3). As a continuous traffic flow model, the Van Aerde formula has the capability to accurately estimate the capacity independent of configurations of the congested and non-congested regimes.

$$d = \frac{1}{h} = \frac{1}{c_1 + \frac{c_2}{s_f - s} + c_3.s} \tag{1}$$

where

d = density (veh/km)

h = distance headway between consecutive vehicles (km)

s_f = free-flow speed (km/h)

c_1 = fixed distance headway parameter (km)

c_2 = first variable headway parameter (km^2/h)

c_3 = second variable distance headway parameter (h^{-1})

s = speed (km/h)

To estimate the model parameters, reasonable starting values for the key traffic flow variables (i.e., capacity, free-flow speed, speed at capacity, and jam density) are assumed and a starting set of parameters (c_1, c_2, c_3, s_f) is calculated. Next, using a nonlinear regression in the speed–density–volume plot, an iterative approach is implemented to estimate the model parameters which minimize the sum of squared errors of the model with respect to dependent variable. Thus, the choice of the dependent variable affects the calibration of the parameters and, as a result, the capacity value. As it is not always clear which variable should be chosen as dependent and which as independent, the orthogonal sum of the squared errors can be minimized as an unbiased compromise using multivariate calibration (12). Once the parameters are calibrated, the capacity can be calculated as the apex volume of the speed–flow diagram.

Stochastic Capacity Estimation Based on Models for Censored Data

The estimation of capacity distribution functions is based on the method proposed by Brilon et al. (14, 15). To estimate the capacity distribution function, traffic breakdowns, that is, the transitions from the uncongested to the congested state, need to be detected. To identify traffic breakdowns, a threshold speed, as the boundary between fluid and congested traffic, is determined by analyzing the speed and flow rate time series. Once the threshold speed is determined, a set of three criteria is applied to detect breakdowns in interval (i) based on 5-minute observations:

1. If the average speed in time interval (i) is above the threshold speed, drops below the threshold speed in the next time interval (i + 1) and remains below for at least 15 minutes (i.e., three consecutive 5-minute intervals), then interval (i) is considered as uncensored, that is, the flow rate in interval (i) represents the momentary capacity of the facility.
2. If the average speed in time interval (i) is above the threshold speed and remains above the threshold speed in the next time interval (i + 1), then interval (i) is considered as censored, that is, the momentary capacity is greater than the observed flow rate.
3. If interval (i) does not satisfy the above conditions, it will not be considered for further analysis.

According to recent applications of the stochastic capacity estimation technique, the capacity distribution function is estimated in 5-minute intervals. In addition, since the *HCM* (*10*) defines the pre-breakdown flow rate as "the 15-minute average flow rate immediately prior to the breakdown event," the capacity distribution is also estimated in 15-minute intervals to receive results comparable with the *HCM*. However, as only short time intervals are appropriate to detect the speed drops, breakdowns are still detected based on 5-minute intervals (as above), but the average of the three consecutive 5-minute flow rates before a breakdown is considered as the uncensored (pre-breakdown) observation. In the same way, the average flow rates of every three 5-minute intervals between a recovery and the following breakdown are considered as censored observations. If the number of 5-minute intervals between the recovery and the breakdown is not an integer multiplier of three, the first one or two 5-minute intervals following the recovery are disregarded.

Once the pre-breakdown (uncensored) and the other non-congested (censored) observations are determined, both non-parametric and parametric approaches can be used to estimate the capacity distribution functions. Equation 2 shows the transformed form of the product limit method (PLM) that is applied to estimate a non-parametric capacity distribution (*14*, *15*).

$$F_c(q) = 1 - S_c(q) = 1 - \prod_{i:q_i<q} \left(\frac{k_i - d_i}{k_i}\right) \quad (2)$$

where

$F_c(q)$ = capacity distribution function
$S_c(q)$ = capacity survival function
q = traffic volume (veh/h)
q_i = traffic volume in interval i (veh/h)
k_i = number of intervals with traffic volume $q_i \leq q$
d_i = number of breakdowns at volume q_i

The maximum likelihood estimation (MLE) technique is employed to estimate the parametric capacity distribution. Here, an a priori assumption about the capacity distribution type is made and the parameters that maximize the log-likelihood value are selected as the calibrated parameters. Equation 3 shows the log-likelihood function that is applied for capacity analysis (*14*, *15*):

$$\ln(L) = \sum_{i=1}^{n}\{\delta_i \cdot ln[f_c(q_i)] + (1-\delta_i) \cdot ln[1 - F_c(q_i)]\} \quad (3)$$

where

$f_c(q_i)$ = density function of capacity
$F_c(q_i)$ = cumulative distribution function of capacity
n = number of intervals
δ_i = 1, if the interval i is uncensored
δ_i = 0, if the interval i is censored

The Weibull distribution, as shown in Equation 4, was suggested by previous research (*8*, *14*, *15*) as the function type that best represents the capacity distribution for freeways:

$$F_c(q) = 1 - e^{-\left(\frac{q}{\beta}\right)^{\alpha}} \quad (4)$$

where

$F_c(q)$ = capacity distribution function
q = traffic volume (veh/h)
α = shape parameter
β = scale parameter (veh/h)

Sustained Flow Index. Once the capacity distribution function is estimated, the SFI can be calculated as the product of the traffic volume (q_i) and the probability of survival at this volume ($S_c(q_i)$). The SFI, which represents the "theoretical average volume that is sustained without a traffic breakdown" (*10*), is given in Equation 5.

$$SFI = q_i \cdot S_c(q_i) = q_i \cdot (1 - F_c(q_i)) \quad (5)$$

where

SFI = sustained flow index (veh/h)
$S_c(q_i)$ = probability of survival at volume q_i
$F_c(q_i)$ = probability of breakdown at volume q_i
q_i = traffic volume in interval i (veh/h)

It is desirable to increase both the probability of survival and the traffic volume for a given freeway section. However, because any increase of volume necessarily leads to a decrease of the survival probability and vice versa, the SFI (as the product of the two) provides a joint performance measure. Thus, the volume that leads to the maximum SFI can be regarded as the best compromise between maximizing the throughput and minimizing the risk of a traffic breakdown. Assuming a Weibull-type capacity distribution, this optimum volume (q_{opt}) is defined in Equation 6.

$$q_{opt} = \beta\left(\frac{1}{\alpha}\right)^{\frac{1}{\alpha}} \tag{6}$$

The SFI and the optimum volume are estimated in both 5- and 15-minute intervals according to the underlying capacity distribution function $F_c(q_i)$.

According to the statistics theory, if a variable is Weibull distributed, its natural logarithm is smallest extreme value (SEV) distributed. Therefore, the natural logarithm of the p-quantile of the Weibull distribution can be written as (*16*):

$$Ln(t_p) = \mu + \sigma(\Phi_{SEV}^{-1}(p)) \tag{7}$$

where

t_p = p-quantile
= time at which the proportion p of the population fails
μ = location parameter = $Ln(\beta)$
σ = scale parameter $= 1/\alpha$
p = probability of failure
$\Phi_{SEV}(z) = 1 - e^{(-e^z)}$ = SEV distribution

Replacing the p-quantile (t_p) in Equation 7 with the optimum volume (q_{opt}) from Equation 6, the probability of breakdown at the optimum volume (P_{opt}) can be calculated as:

$$P_{opt} = 1 - e^{-\frac{1}{\alpha}} \tag{8}$$

As can be seen in Equation 8, the breakdown probability at the optimum volume depends only on the value of the shape parameter of the Weibull distribution. As the shape parameter increases, the probability of breakdown at the optimum volume decreases. Therefore, control strategies such as variable speed limits that increase the shape parameter (i.e., reduce variance of the distribution function, cf. (*17*)) decrease the probability of breakdown.

Confidence Intervals. Estimated values for the parameters of the capacity distribution function and the optimum volume strongly depend on the collected data sample in the sense that another set of parameters will be estimated if another sample is collected. Thus, it is important to estimate the confidence intervals for these parameters to address sample to sample variations.

Assuming a Weibull distribution for the capacity of a freeway, Wald confidence intervals can be estimated for both the parameters of the capacity distribution function and the optimum volume. However, if another assumption is made for the capacity distribution function (e.g., normal distribution), the optimum volume may only be derived with numeric calculations and the Wald confidence intervals cannot be easily computed. Thus, this research provides Bootstrap confidence intervals, which are computed numerically and widely used in many statistical analyses, to compare the proximity of the estimation results for future reference in case only the Bootstrap confidence intervals are applicable.

Wald Confidence Intervals. Once the scale parameter of the Weibull-type capacity distribution function is calibrated, its standard error can be easily estimated with the Taylor series approximation and the Wald confidence interval is calculated according to Equation 9. The same method can be applied to estimate the Wald confidence interval for the shape parameter:

$$\hat{\beta} \pm z_{(1-\alpha/2)}\widehat{se}_{\hat{\beta}} \tag{9}$$

where

$\hat{\beta}$ = scale parameter (veh/h)
$\widehat{se}_{\hat{\beta}}$ = standard error the scale parameter
$z_{(1-\alpha/2)}$ = standard score at $\alpha\%$ significance level

To estimate the Wald confidence interval for the optimum volume, a hand calculation is needed to estimate its standard error. Thus, if the p-quantile (t_p) in Equation 7 is replaced with the optimum volume (q_{opt}) from Equation 6, its variance can be estimated with Equation 10 using the Taylor series approximation.

$$\widehat{Var}_{\widehat{Ln(q_{opt})}} = \left[\frac{\partial Ln(q_{opt})}{\partial \mu}\right]^2 \cdot \widehat{Var}(\hat{\mu}) + \left[\frac{\partial Ln(q_{opt})}{\partial \sigma}\right]^2 \cdot \widehat{Var}(\hat{\sigma}) + 2 \cdot \frac{\partial Ln(q_{opt})}{\partial \mu} \cdot \frac{\partial Ln(q_{opt})}{\partial \sigma} \cdot \widehat{Cov}(\hat{\mu}, \hat{\sigma}) \tag{10}$$

Since $\frac{\partial Ln(q_{opt})}{\partial \hat{\mu}} = 1$, $\frac{\partial Ln(q_{opt})}{\partial \hat{\sigma}} = -Ln(\alpha)$, $\widehat{VAR}(\hat{\mu})$, $\widehat{VAR}(\hat{\sigma})$, and $\widehat{COV}(\hat{\mu}, \hat{\sigma})$ are provided by the software, the variance and, as a result, the standard error of the natural logarithm of the optimum volume ($Ln(q_{opt})$) can be calculated. Thus, the confidence interval for the natural logarithm of the optimum volume can be estimated with Equation 11.

$$\widehat{Ln(q_{opt})} \pm z_{(1-\alpha/2)}\widehat{se}_{\widehat{Ln(q_{opt})}} \tag{11}$$

Next, an antilog can be taken from the upper and lower bounds of Equation 11 to estimate the Wald confidence interval for the optimum volume (q_{opt}).

Bootstrap Confidence Intervals. Bootstrap confidence intervals are approximate confidence intervals that are built based on sampling with replacement. In this method, once the sample data (of size n) is collected, B Bootstrap samples of the same size (also called resamples) are taken from the initial sample with replacement. This procedure suggests that in a single Bootstrap sample some observations may be repeated whereas some others may not be selected even once. Thus, new parameters are estimated for each of the Bootstrap samples. If a sufficient number (usually between 2,000 and 5,000) of Bootstrap samples is selected, the confidence intervals delivered for the parameters by this method are quite similar to those delivered by exact methods. Also, Bootstrap confidence intervals have the advantage of relaxing the assumption of the underlying distribution of the data that is used to build the conventional confidence intervals and, as a result, are used in many statistical analyses (*16*).

Empirical Results

To compare the capacity values delivered with the conventional and stochastic methods discussed above, an empirical analysis was performed based on traffic flow data in 5-minute intervals collected from 19 urban freeway bottleneck sections located in California. All sections under study were merge bottlenecks with a considerable share of commuter traffic. To avoid that spillbacks from downstream affect the capacity estimation, only distinct bottlenecks were selected by analyzing long-time speed contour plots. A same sample size of one year was used to estimate both deterministic and stochastic capacities for every section under investigation. With this large sample size, a sufficient number of traffic breakdowns could be observed at each bottleneck section in order to estimate a reliable capacity distribution function. To consider the influence of different prevailing trip purposes on the capacity, only workdays were considered for analysis. Also, to assure that non-recurrent congestions (such as those caused by accidents or incidents) did not affect the estimated capacity values, traffic breakdowns at volumes less than 1,200 veh/h/lane were not considered for analysis (*18*).

In order to calculate the conventional capacity values, parameters of the Van Aerde model were estimated by applying multivariate calibration. Thus, the SPD_CAL software was used to minimize the orthogonal sum of squared errors of speed, volume, and density using non-linear regression (*19*, *20*). The capacity obtained as the volume at the apex of the Van Aerde model is referred to as c_{VAM} in the following.

To calculate the optimum volumes by maximizing the SFI, both non-parametric and parametric capacity distribution functions were estimated. To calibrate the parametric distribution function, the Weibull distribution was assumed to represent the capacity. Once the parameters were calibrated, optimum volumes and their corresponding breakdown probabilities were calculated according to Equations 6 and 8, respectively. Figure 1 shows the capacity distribution functions and the SFI curves estimated based on 5-minute and 15-minute intervals for a bottleneck section located at Interstate 5. As can be seen, for 5-minute intervals, the freeway section reaches its maximum SFI at a volume of 7,630 veh/h. This optimum volume corresponds to a 4.6 and 15.6% probability of breakdown for the distribution functions estimated in 5- and 15-minute intervals, respectively. Based on 15-minute data, an optimum volume of 7,185 veh/h was estimated, which corresponds to a 4.5% probability of breakdown for the capacity distribution function estimated in 15-minute intervals.

For better illustration, Figure 2 shows the SFI superimposed over the Van Aerde model for the same four-lane freeway section. Here, the conventional capacity estimated in 5-minute intervals by multivariate calibration of the Van Aerde model is 7,472 veh/h. Once both the conventional capacity values and the optimum volumes are estimated, they can be compared based on the percentage difference between the results delivered by the two methods. For the section shown in Figure 2, the optimum volume is 2% greater than the conventional capacity estimate.

Table 1 shows the results of applying the above procedure for all sections under investigation. As can be seen, the average probability of breakdown corresponding to the 5-minute optimum volume is 4.5 and 16.4% for the 5- and 15-minute capacity distribution functions, respectively. The 16.4% probability of breakdown found in this research is roughly consistent with the 15% probability of breakdown recommended by the *HCM* 6th *edition* (*10*) to select a single value from the 15-minute capacity distribution function. However, if the optimum volume is estimated based on 15-minute intervals, the average probability of breakdown in 15-minute intervals corresponding to this 15-minute optimum volume is 3.9%.

The average probability of breakdown corresponding to the conventional capacity values c_{VAM} estimated with the Van Aerde model is 3.7% for the 5-minute capacity distribution function. This percentage is similar to the 4% probability of breakdown obtained by Modi et al (*8*) for U.S. freeways. The results also reveal that the standard deviation of the probability of breakdown at the optimum

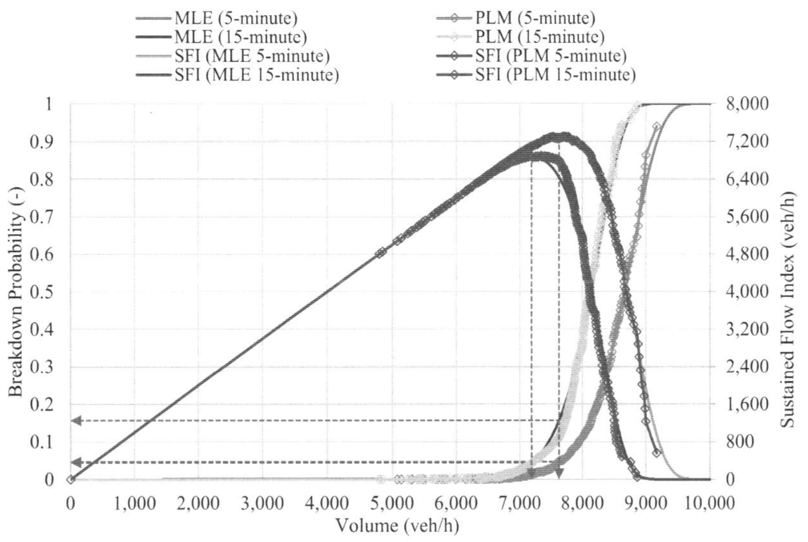

Figure 1. Estimated capacity distribution functions obtained with the product limit method (PLM) and the maximum likelihood estimation technique (MLE) and corresponding sustained flow indexes (SFI) based on 5- and 15-minute intervals for a four-lane freeway cross-section.

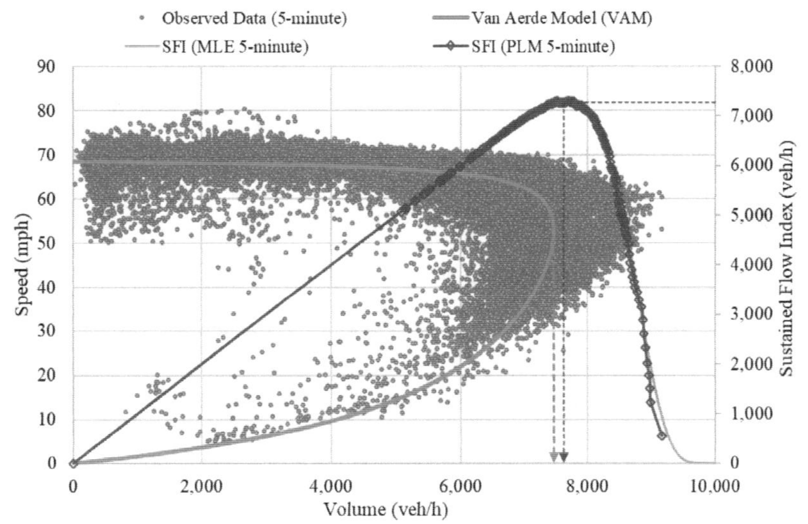

Figure 2. Conventional capacity vs. optimum volume estimated in 5-minute intervals for a four-lane freeway cross-section.

volume is less than that of the conventional capacity: $stdev(F_{c,5}(q_{opt})) = 0.42 < stdev(F_{c,5}(c_{VAM})) = 1.22$. This indicates that the probability of breakdown at the optimum volume varies less than the probability of breakdown corresponding to the volume at the apex of the Van Aerde curve.

The average percentage difference between the optimum volumes and the capacities estimated with the Van Aerde model is 1.2% and the correlation between them is more than 99%. These findings suggest that the optimum volumes obtained by maximizing the SFI are on average within a reasonable range of the conventional

Table 1. Estimated Conventional Capacity Values and Optimum Volumes Based on 5- and 15-minute Intervals as Well as Their Corresponding Probabilities of Breakdown for the Freeway Sections under Investigation

Section	Lanes	Capacity $c_{VAM,5}$ (veh/h)	$F_{c,5}(c_{VAM,5})$	Optimum volume $q_{opt,5}$ (veh/h)	$F_{c,5}(q_{opt,5})$	$F_{c,15}(q_{opt,5})$	Optimum volume $q_{opt,15}$ (veh/h)	$F_{c,15}(q_{opt,15})$
1	2	3,516	2.9	3,611	4.8	20.5	3,363	4.1
2	2	3,942	5.8	3,899	4.7	18.5	3,653	3.9
3	2	3,762	2.3	3,880	4.4	15.3	3,670	3.6
4	2	3,735	2.7	3,815	4.4	15.3	3,611	3.5
5	3	5,574	3.6	5,639	4.6	20	5,253	4.2
6	3	5,363	6.6	5,286	5.1	16.6	4,977	3.9
7	3	5,696	3.4	5,800	4.9	16.8	5,450	4.2
8	3	5,994	4.4	5,990	4.3	14.5	5,678	3.7
9	4	8,306	4.1	8,254	3.4	11.6	7,921	2.7
10	4	8,378	4.2	8,393	4.4	14.4	7,960	3.6
11	4	7,486	4.3	7,494	4.4	12.6	7,138	4.1
12	4	8,416	2.9	8,635	4.8	17.5	8,085	4.5
13	4	7,472	3	7,630	4.6	15.6	7,185	4.5
14	4	7,684	2.5	7,841	4.1	15.8	7,596	3.6
15	4	7,154	2.6	7,273	3.9	14.6	6,897	3.5
16	4	7,826	2.7	8,098	5.1	19.2	7,543	4.6
17	5	9,249	5.5	9,151	4.3	18.4	8,550	4.4
18	5	8,001	2.8	8,155	4.3	21.3	7,591	3.9
19	5	9,995	4.1	10,004	4.2	13.3	9,509	3.9
Average prob. of breakdown			3.7		4.5	16.4		3.9

capacity estimates, and can therefore be considered as suitable capacity estimates.

As can be seen in Table 1, the estimation results suggest that there is no evident empirical relationship between the number of lanes and the probability of breakdown at the optimum volume for the distribution functions estimated based on 5- and 15-minute intervals. This seems reasonable as, according to Equation 8, the probability of breakdown at the optimum volume only depends on the shape parameter of the Weibull distribution and previous research did not find a meaningful relationship between the number of lanes and the shape parameter either. Therefore, it can be concluded that the probability of breakdown at the optimum volume for freeways with different number of lanes is nearly constant.

Based on the calibrated parameters of the Weibull distribution for 5-minute intervals, Bootstrap confidence intervals were calculated. For this, 10,000 Bootstrap samples were drawn (with replacement) from the initial sample used to estimate parameters of the distribution function. For each of the Bootstrap samples, a unique set of parameters and, as a result, the optimum volumes were estimated.

After estimating the parameters and the optimum volumes, their confidence intervals at a significance level of 5% were calculated. To estimate the Wald confidence intervals for the estimated parameters and optimum volumes, Equations 9 and 11, respectively, were applied. Both Bootstrap-based and Wald confidence intervals estimated at a significance level of 5% are shown in Table 2. As can be seen, the results delivered by both methods are very similar, which means that applying the Bootstrapping technique delivers a suitable approximation of confidence intervals for the optimum volume in case that another function type is assumed for the capacity distribution (whose optimum volume can only be estimated numerically). Also, since the estimated 95% confidence intervals are relatively small, it can be inferred that the optimum volume is a fairly stable indicator of freeway capacity.

The formulation of the optimum volume and its correspondence with conventional capacity (i.e., $q_{opt} = \beta(1/\alpha)^{1/\alpha}$ and $c \approx q_{opt}$ thus $c \approx \beta(1/\alpha)^{1/\alpha}$) implies that a relationship exists between the shape and scale parameters of the capacity distribution function for any capacity value. Thus, as shown in Figure 3, different combinations of the shape and the scale parameters may lead to the same capacity value. As the relationship between the shape and scale parameter is true for any capacity value, a set of indifference curves (one for each capacity value) was created.

Given a conventional capacity value, for example the design capacity obtained from the *HCM* (*10*), and a reasonable assumption for the shape parameter, the scale parameter can be calculated and the capacity distribution function can be estimated. For standard conditions, practitioners may assume a shape parameter of $\alpha = 22$

Table 2. Estimated Wald and Bootstrap Confidence Intervals for the Parameters and Optimum Volumes

Section number		1	2	3	4	5	6	7	8	9	10	11	12	13	14	15	16	17	18	19
Shape parameter		20.2	20.7	22.0	22.5	21.3	19.2	20.1	22.8	28.6	22.2	22.4	20.1	21.1	23.9	25.1	19.0	22.6	22.7	23.1
Wald CI	Lower bound	19.4	19.6	20.9	21.1	19.2	17.9	18.7	21.4	26.9	20.5	21.5	19.4	20.0	22.2	23.4	18.0	21.9	21.1	21.7
	Upper bound	21.0	21.8	23.2	24.0	23.5	20.6	21.6	24.4	30.6	24.4	23.4	21.0	22.2	26.0	27.2	20.1	23.4	24.7	24.7
Bootstrap CI	Lower bound	19.2	19.7	20.5	21.0	19.2	17.9	18.5	21.0	26.0	19.9	21.4	19.1	19.9	22.0	22.8	17.4	21.4	21.0	20.9
	Upper bound	21.2	21.7	23.7	24.1	23.8	20.8	21.8	24.8	32.3	24.8	23.5	21.2	22.3	26.0	27.9	21.1	23.9	24.8	25.5
Scale parameter		4,190	4,515	4,465	4,382	6,509	6,164	6,735	6,871	9,281	9,649	8,609	10,023	8,818	8,954	8,269	9,457	10,504	9,357	11,460
Wald CI	Lower bound	4,165	4,481	4,435	4,349	6,419	6,110	6,661	6,809	9,226	9,548	8,567	9,961	8,753	8,882	8,211	9,361	10,463	9,288	11,381
	Upper bound	4,216	4,549	4,496	4,415	6,621	6,227	6,811	6,934	9,336	9,751	8,652	10,087	8,883	9,026	8,327	9,554	10,545	9,426	11,539
Bootstrap CI	Lower bound	4,162	4,482	4,427	4,346	6,414	6,104	6,655	6,796	9,187	9,543	8,564	9,947	8,756	8,881	8,205	9,279	10,447	9,289	11,364
	Upper bound	4,221	4,550	4,508	4,421	6,626	6,232	6,827	6,953	9,373	9,780	8,657	10,108	8,888	9,039	8,343	9,623	10,573	9,429	11,578
Optimum volume		3,611	3,899	3,880	3,815	5,639	5,286	5,800	5,990	8,254	8,393	7,494	8,635	7,630	7,841	7,273	8,098	9,151	8,155	10,004
Wald CI	Lower bound	3,598	3,880	3,861	3,793	5,593	5,252	5,760	5,959	8,211	8,330	7,463	8,598	7,589	7,791	7,223	8,056	9,118	8,082	9,950
	Upper bound	3,624	3,918	3,898	3,837	5,686	5,321	5,841	6,021	8,298	8,457	7,526	8,673	7,671	7,891	7,324	8,140	9,184	8,228	10,059
Bootstrap CI	Lower bound	3,598	3,881	3,863	3,797	5,594	5,257	5,761	5,958	8,221	8,338	7,468	8,596	7,594	7,801	7,235	8,024	9,123	8,102	9,962
	Upper bound	3,625	3,919	3,898	3,835	5,696	5,319	5,846	6,027	8,293	8,448	7,521	8,676	7,669	7,881	7,316	8,175	9,179	8,213	10,050

(according to the average value given in Table 2). However, given the fact that the shape parameter is proportional to the variance of the capacity distribution function, in case it is believed that the capacity variance is lower or higher than for standard conditions, a smaller or greater shape parameter may be selected. For example, it seems reasonable to assume a greater shape parameter if control strategies such as variable speed limits that reduce the capacity variance (cf. [*17*]) are implemented in the segment under investigation. This method allows practitioners to estimate a capacity distribution function for segments whose conventional capacity has already been estimated or can be taken from the guidelines in case that an empirical estimation of the capacity distribution function is not feasible.

Conclusions

The stochastic variability of freeway capacity has evolved to become a widely accepted concept in traffic engineering. Different methods have been proposed and applied to estimate the capacity distribution function. Although traffic flow assessment procedures given in guidelines like the *HCM* (*10*) are still mostly based on conventional (deterministic) design capacities, only limited research has been carried out on how to select a single, representative, capacity value from the estimated capacity distribution function. A recent study by Shojaat et al. (*9*) introduced the SFI as a new performance measure based on the stochastic notion of the capacity and suggested selecting the volume that maximizes the SFI (referred to as optimum volume) as the capacity of the freeway section. However, so far these values were not compared with conventional capacity estimates to assess their validity. To verify the reasonableness of the results obtained by maximizing the SFI, in this study, optimum volumes were compared with the conventional capacities based on data collected for 19 U.S. freeway bottleneck sections. Despite the application of two completely different methodologies, the obtained average differences between the estimated capacity values are small (nearly 1.2%), and a very strong correlation between the capacity estimates exists.

The empirical results revealed that the optimum volume estimated in 5-minute intervals is roughly equivalent to a 5% probability of breakdown in 5-minute intervals and a 15% probability of breakdown in 15-minute intervals, respectively. Therefore, the optimum volume obtained by maximizing the SFI in 5-minute intervals well corresponds with the 15-minute capacity according to the *HCM* (*10*), which proposes a 15% probability of breakdown for the capacity estimation from field data. Mathematical derivations also yielded that the probability of breakdown at the optimum volume solely depends

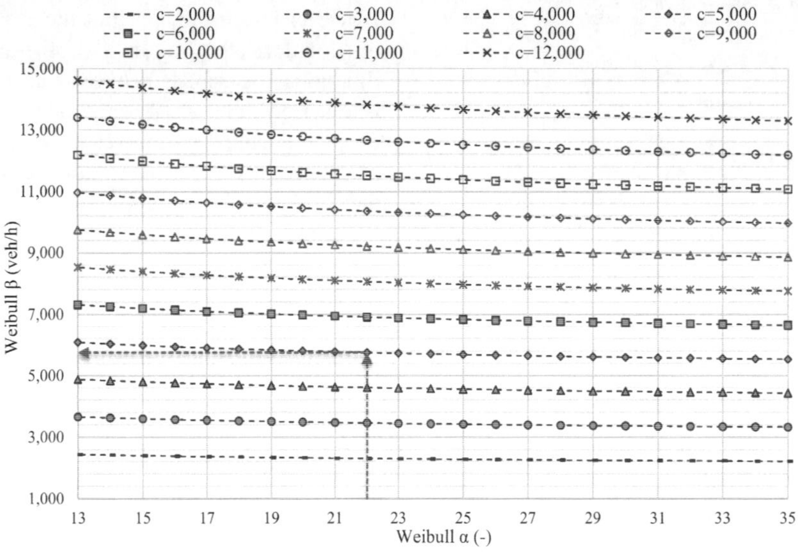

Figure 3. Indifference curves for different capacity values.

on the shape parameter of the Weibull-type capacity distribution and thus can be expected to be relatively constant for freeways with different numbers of lanes.

To investigate the stability of the estimated optimum volumes, a new procedure was developed to calculate the 95% confidence intervals with the Wald and Bootstrap methods. The results obtained by the two methods are very similar, and suggest that the confidence intervals of the optimum volumes are relatively small. Therefore, it can be concluded that the optimum volume is a fairly stable indicator of freeway capacity. Furthermore, based on the correspondence between the optimum volume and the conventional capacity, a practical method to estimate the entire capacity distribution function given a conventional capacity estimate was developed.

Based on the presented findings, it is suggested that the optimum volume obtained by maximizing the SFI may be considered as a preferred approach to select a single capacity value from the capacity distribution function. Compared with the conventional approach of estimating the capacity in the speed–flow diagram, the optimum volume solely relies on uncongested flow observations and therefore best represents the pre-breakdown capacity. The application of the proposed approach is simple, as the optimum volume can be calculated from the parameters of the capacity distribution function by a single equation.

Acknowledgments

This research was funded by the United States Department of Transportation through its University Transportation Center (UTC) program at the Gulf Coast Center for Evacuation and Transportation resiliency, a member of the National Transportation Center at the University of Maryland. The data-set used in this research was collected from the Caltrans Performance Measurement System (PeMS) website.

Author Contributions

The authors confirm contribution to the paper as follows: study conception and design: Siavash Shojaat, Brian Wolshon, Justin Geistefeldt, Luis Escobar; data collection: Siavash Shojaat; analysis and interpretation of results: Siavash Shojaat, Justin Geistefeldt, Luis Escobar, Scott Parr; draft manuscript preparation: Siavash Shojaat, Brian Wolshon, Justin Geistefeldt, Scott Parr. All authors reviewed the results and approved the final version of the manuscript.

References

1. Hall, F. L., and K. Agyemang-Duah. Freeway Capacity Drop and the Definition of Capacity. *Transportation Research Record: Journal of the Transportation Research Board*, 1991. 1320: 20–28.
2. Banks, J. H. Flow Processes at a Freeway Bottleneck. *Transportation Research Record: Journal of the Transportation Research Board*, 1991. 1278: 20–28.
3. Persaud, B., S. Yagar, and R. Brownlee. Exploration of the Breakdown Phenomenon in Freeway Traffic. *Transportation Research Record: Journal of the Transportation Research Board*, 1998. 1643: 64–69.
4. Lorenz, M., and L. Elefteriadou. *Transportation Research Circular E-C018: A Probabilistic Approach to Defining Freeway Capacity and Breakdown*. TRB, National Research Council, Washington, D.C., 2000. http://onlinepubs.trb.org/onlinepubs/circulars/ec018/ec018toc.pdf

5. Elefteriadou, L., and P. Lertworawanich. Defining, Measuring and Estimating Freeway Capacity. Presented at 82nd Annual Meeting of the Transportation Research Board, Washington, D.C., 2003.
6. Minderhoud, M. M., H. Botma, and P. H. L. Bovy. Assessment of Roadway Capacity Estimation Methods. *Transportation Research Record: Journal of the Transportation Research Board*, 1997. 1572: 59–67.
7. Geistefeldt, J. *Transportation Research Circular E-C149: Empirical Relation Between Stochastic Capacities and Capacities Obtained from the Speed-Flow Diagram*. Transportation Research Board of the National Academies, Washington, D.C., 2011. http://onlinepubs.trb.org/onlinepubs/circulars/ec149.pdf
8. Modi, V., A. Kondyli, S. Washburn, and D. S. McLeod. Freeway Capacity Estimation Method for Planning Applications. *Journal of Transportation Engineering (ASCE)*, No. 140, 2014.
9. Shojaat, S., J. Geistefeldt, S. A. Parr, C. G. Wilmot, and B. Wolshon. Sustained Flow Index: A Stochastic Measure of Freeway Performance. *Transportation Research Record: Journal of the Transportation Research Board*, 2016. 2554: 158–165.
10. *Highway Capacity Manual: A Guide for Multimodal Mobility Analysis*, 6th ed. Transportation Research Board, Washington, D.C., 2016.
11. Van Aerde, M. A Single Regime Speed-Flow-Density Relationship for Freeways and Arterials. Presented at 74th Annual Meeting of the Transportation Research Board, Washington, D.C., 1995.
12. Van Aerde, M., and H. Rakha. Multivariate Calibration of Single Regime Speed-Flow-Density Relationships. *Proceedings of the Vehicle Navigation and Information Systems (VNIS) conference*, Seattle, Washington, 1995.
13. Van Toorenburg, J. A. C. *Praktijkwaarden voor de Capaciteit (Practical Values for Capacity)*. Transport Research Centre (DKV), Ministry of Transport, Public Works and Water Management, Rotterdam, 1986.
14. Brilon, W., J. Geistefeldt, and M. Regler. Reliability of Freeway Traffic Flow: A Stochastic Concept of Capacity. *Proc., 16th International Symposium on Transportation and Traffic Theory*, Elsevier Ltd., Oxford, 2005, pp. 125–144.
15. Brilon, W., J. Geistefeldt, and H. Zurlinden. Implementing the Concept of Reliability for Highway Capacity Analysis. *Transportation Research Record: Journal of the Transportation Research Board*, 2007. 2027: 1–8.
16. Meeker, W. Q., G. J. Hahn, and L. A. Escobar. *Statistical Intervals: A Guide for Practitioners and Researchers*, 2nd ed. John Wiley & Sons, Inc., New York, 2017.
17. Geistefeldt, J. Capacity Effects of Variable Speed Limits on German Freeways. *Procedia Social and Behavioral Sciences*, No. 16, 2011, pp. 48–56.
18. Geistefeldt, J., and W. Brilon. A Comparative Assessment of Stochastic Capacity Estimation Methods. *Transportation and Traffic Theory 2009: Golden Jubilee, Proc., 18th International Symposium on Transportation and Traffic Theory*, Springer, Dordrecht, Heidelberg, London, New York, 2009, pp. 583–602.
19. Rakha, H., and M. Arafeh. Calibrating Steady-State Traffic Stream and Car-Following Models Using Loop Detector Data. *Transportation Science*, Vol. 44, No. 2, 2010, pp. 151–168.
20. Rakha, H. *Traffic Stream Calibration Software*. Virginia Polytechnic Institute. https://sites.google.com/a/vt.edu/hrakha/software. Accessed January 23, 2017.

The Standing Committee on Highway Capacity and Quality of Service (AHB40) peer-reviewed this paper (18-02360) and selected it to receive its 2018 Best Paper Award.

Article

TRR
JOURNAL OF THE TRANSPORTATION RESEARCH BOARD

Transportation Research Record
2018, Vol. 2672(15) 142–154

DOI: 10.1177/0361198118776112
journals.sagepub.com/home/trr
SAGE

Multimodal Level of Service Methodologies: Evaluation of the Multimodal Performance of Arterial Corridors

Natalia Zuniga-Garcia[1], Heidi W. Ross[2], and Randy B. Machemehl[1]

Abstract

The principal objective of this research is to evaluate the multimodal performance of arterial corridors using currently available multimodal level of service (MMLOS) methodologies. Eight different MMLOS approaches are applied to a case study using an arterial corridor section in Austin, Texas. The methodologies applied are: *Highway Capacity Manual; Transit Capacity and Quality of Service Manual*; Charlotte, NC, *Urban Street Design Guidelines*; pedestrian and bicycle environmental quality indices; assessment of level of traffic stress; bicycle compatibility index; deficiency index; and Walk Score®, Bike Score®, and Transit Score®. The analysis is focused on the assessment of pedestrian, bicycle, and transit modes. The methodologies are evaluated and contrasted. The paper provides a comprehensive review of the current state of practice of multimodal evaluation and recommendations about the most appropriate approaches to assess multimodal performance of arterial corridors.

The idea that streets are places— not just paths to and through places—has increasingly guided transportation thinking and investment over the past decade (*1*). The Transportation Equity Act for the 21st Century (TEA-21) and its predecessor, the Intermodal Surface Transportation Efficiency Act of 1991 (ISTEA), called for mainstreaming transit, pedestrian, and bicycle projects into the planning, design, and operation of the U.S. transportation system (*2*). With an increase in transportation funding, the need arises to implement analytic tools that help measure the effects of investments and guide decisions regarding the effective planning, design, and operation of streets in multimodal environments.

The most common evaluation approach is to use qualitative measures that relate to the quality of service, also known as the level of service (LOS). Traditionally, the LOS metric has been focused on assessing automobile-oriented measures. The LOS assigns a letter from A to F based on the quality of transportation as defined by various performance measurements, such as speed, volume-to-capacity ratio, and delay, among others. The letter A corresponds to the best quality of service, while the letter F represents the most deficient condition.

The concept of LOS has been expanded beyond automobiles to include the quality of service as perceived by cyclists, pedestrians, and transit users. This is known as multimodal level of service (MMLOS). While there are several different MMLOS metric approaches, there is no generally accepted method for combining auto, transit, bicycle, and pedestrian modes into a unique performance measure. This could be a limitation for planning projects that require an overall estimation for an entire section.

The principal objective of this research is to evaluate the multimodal performance of arterial corridors using currently available MMLOS methodologies and to compare and contrast results obtained by using the methodologies. To date, it has not been common to develop a single LOS to characterize all modes in a corridor. This research further assesses the reasonableness of doing so. The analysis includes a case study of an arterial corridor section located in Austin, Texas.

The main contributions of this paper include: (a) a comprehensive review of the available MMLOS methodologies; (b) an evaluation and contrasting of the MMLOS approaches using a case study; and (c) insights about the most appropriate multimodal evaluation procedure for arterial corridors. The subsequent sections of

[1]Department of Civil, Architectural and Environmental Engineering, The University of Texas at Austin, Austin, TX
[2]Center for Transportation Research, The University of Texas at Austin, Austin, TX

Corresponding Author:
Address correspondence to Natalia Zuniga-Garcia: nzuniga@utexas.edu

this paper are: "Measuring the Multimodal Performance," which presents the state of practice of the MMLOS methods; "Case Study," which provides details about the case study; and "Results and Discussion," which presents the results and analysis of the application of the selected methods to the case study. The final section, "Summary and Conclusions," summarizes the main findings.

Measuring the Multimodal Performance

The most prominent and widely used method to measure the performance of roadways is presented in the *Highway Capacity Manual* (*HCM*), which first introduced the concept of LOS in the second edition (1965). However, it was not until the fifth edition in 2010 when *HCM* included a multimodal analysis framework. This was the first *HCM* to provide an integrated multimodal approach to the analysis and evaluation of urban streets from the point of view of automobile drivers, transit passengers, cyclists, and pedestrians (*3*). The most recent *HCM*, published in 2016, notably adds the subtitle *A Guide for Multimodal Mobility Analysis* and "underscore[s] the focus on evaluating the operational performance of several modes including pedestrian and bicycles, and their interactions" (*4*).

The main concern about the *HCM* editions prior 2010 is that the LOS reflected a motorist perspective with the emphasis on automobile traffic, density, delay, and speed. Additionally, there is a lack of interaction between automobile, pedestrian, transit, and bicycle LOS metrics (*2*). For instance, the pedestrian LOS criteria for sidewalks is based on facility width and pedestrian volume. This concept is derived from the automobile LOS methodology where density plays a critical role. However, pedestrians may be more interested in other aspects such as the presence of buffers or the volume and speed of adjacent automobiles.

In response to the limitations found in the *HCM* approach before the 2010 edition, and due to the increasing interest in evaluating multimodal scenarios, several scholars and local jurisdictions have developed their own MMLOS methodologies (*5*). The methodologies can be classified into two groups. The first group includes methods that employ the traditional letter-grade (A to F) LOS metric. The second group consists of methods that use various score methodologies or checklists to evaluate multimodal characteristics

Level of Service Metric

Highway Capacity Manual. The research base for the *HCM* MMLOS framework comes from the National Cooperative Highway Research Program (NCHRP) Project NCHRP 3-70. The NCHRP Report 616 (*2*) obtained as part of this project summarizes the methodology used to develop the MMLOS framework. The study used as inputs models first developed by Bruce Landis and Theo Petritsch (*6–8*) in an effort by the Florida Department of Transportation (FDOT) to develop its own multimodal performance measures. The *HCM* uses four units of analysis: intersections, links, segments, and facility. The LOS estimation requires information about demand, control, and geometry. The equations provide a numerical score that is converted into a letter based on five ranges: A ($\leq$2.00), B (2.00–2.75), C (2.75–3.50), D (3.50–4.25), E (4.25–5.00), and F (>5). The LOS results are provided for the different modes, and the procedure does not provide a single LOS measure that combines all of them.

Transit Capacity and Quality of Service Manual. The *Transit Capacity and Quality of Service Manual* (*TCQSM*) (*9*) is the transit counterpart to the *HCM*. The manual contains background, statistics, and graphics on various types of public transportation, and it provides a framework for measuring transit availability, comfort, and convenience from a passenger point of view. The manual contains quantitative techniques for calculating the capacity of bus, rail, and ferry transit services and transit stops, stations, and terminals (*4*). The latest version is the third edition, published in 2013.

The *TCQSM* method distinguishes between demand-responsive transit and fixed-route transit services. The analysis is applied at a street-segment level. The manual presents a two-dimensional LOS framework covering two service quality dimensions: availability, and comfort and convenience. The availability dimension considers frequency, service span, and access. The comfort and convenience dimension core measures are passenger load, reliability, and travel time.

FDOT Quality/Level of Service Handbook. The *Quality/Level of Service Handbook* (*Q/LOS*), developed by FDOT, provides LOS measures, thresholds, and estimation methodologies for automobile, transit, bicycle, and pedestrian modes. For more than 15 years FDOT has been a national leader in research related to developing MMLOS metrics (*10*).

FDOT first developed the *Q/LOS* method, and subsequently led and supported extensive research on MMLOS that helped to inform the inclusion of MMLOS in the *HCM*'s fifth (2010) and sixth (2016) editions (*10*). The most recent version of *Q/LOS* was published in 2013 and includes new analytical techniques contained in *HCM* 2010 (*11*). The handbook provides a description of the simplifying assumptions and planning extensions FDOT has made to the *HCM*. Additionally, it specifies six analysis techniques used only by FDOT: generalized service volume tables (recently included in the *HCM Applications Guide* (*12*)), freeway facility capacities, rural freeway LOS criteria, arterial facility LOS criteria for automobiles, arterial free-flow speed, and passing lanes on two-lane highways.

Charlotte, NC: Urban Street Design Guidelines. The City Council of Charlotte, North Carolina, adopted *Urban Street Design Guidelines* (*USDG*) in 2007 (*13*) as a tool for planning and designing "complete streets." In 2009 these guidelines received the U.S. Environmental Protection Agency Award for Smart Growth Achievement (*14*). *USDG* methodology identifies and evaluates features according to their influence on the comfort and safety of pedestrians and cyclists.

USDG is focused on pedestrian and bicycle LOS and its evaluation methodology consists of adding or subtracting points for the presence or absence of particular features. This method differs from the *HCM, TCQSM*, and *Q/LOS* methodologies, where a numeric score is obtained using formulas produced through linear regression. Finally, scores (ranging from 0 to 100) are compared with limits that determine the A to F grades for LOS for each mode. The main limitation of this method is that it only applies to intersections, not to street segments. Additionally, there are concerns about the metric's lack of system-user validation, transparency in selection and development of variables, and weights for the points system (*2*).

City of Fort Collins Multimodal Transportation Level of Service Manual. Fort Collins, Colorado, implemented a *Multimodal Transportation Level of Service Manual* in 1997 as part of its "Community Vision and Goals 2015" (*15*). The manual provides LOS for automobile, pedestrian, cyclist, and transit. The automobile LOS is estimated using the *HCM* procedure (*16*). The bicycle LOS standards are based on access to various facilities within connecting corridors. The analysis only includes bicycle facilities that are consistent with the city's minimum design standards, which is the principal limitation of the method. The pedestrian LOS methodology defines minimum LOS requirements according to the five types of areas established by the city: pedestrian district, activity corridor/center, transit corridor, school walking area, and other. The methodology considers pedestrian characteristics that are not included in other methods, such as continuity, security, and amenities, which are important from a pedestrian planning point of view. The transit LOS standards consider route service characteristics, and land-use features of the area served.

Bicycle Compatibility Index. The Bicycle Compatibility Index (BCI) (*17*) was developed to evaluate the capability of urban and suburban roadway sections to accommodate both motorists and cyclists using geometric and operational characteristics such as lane widths, speed, and volume. It was possible to develop a linear regression model to predict the BCI using the results of a Video Laboratory survey. This model predicts a cyclist's overall comfort-level rating using significant variables and adjustment factors. The BCI obtained using the model is compared with six compatibility levels: A, extremely high; B, very high; C, moderately high; D, moderately low; E, very low; and F, extremely low. It is important to note that the BCI model is for mid-block street segments only. The ratings do not account for major intersections along the route where the cyclist may encounter a stop sign or traffic signal (*17*).

Modal Score Measures

Pedestrian and Bicycle Environmental Quality Index. In 2007, the San Francisco Department of Public Health developed the Pedestrian Environmental Quality Index (PEQI) (*18*) and the Bicycle Environmental Quality Index (BEQI) (*19*). The indices are observational surveys that quantify streets and intersection factors that affect pedestrians and cyclists, and evaluate what streetscape improvements could be made to promote walking and cycling. Both indices are organized into five categories: intersection safety, traffic, street design, land use, and perceived safety.

The methodology used to estimate the PEQI and BEQI scores is similar to the Charlotte *USDG*, where the user adds or subtracts points according to the indicators present in the section analyzed. Each index of the five principal categories has a scaling factor or weight. An expert panel was convened to assign weights to each attribute that reflect the level of importance each indicator has for pedestrians when evaluating street segments and intersections (*20*). Both PEQI and BEQI are scored on a scale from 0 to 100, where 100 is the most desirable. The indicators can be aggregated to create the final index, which can be reported as an overall index score or deconstructed by categories.

Level of Traffic Stress. Researchers at the Mineta Transportation Institute in San José, California, developed the level of traffic stress (LTS) method (*21*) to evaluate the LOS for bicycle travel. The authors classified bicycle facilities into four groups. LTS 1 is meant to be a level that most children can tolerate, while LTS 2 is the level that will be tolerated by the typical adult population. LTS 3 is the level tolerated by cyclists who are "enthused and confident," but still prefer having their own dedicated space for riding. LTS 4 is the level tolerated only by those characterized as "strong and fearless."

The LTS methodology is applied at intersection and street-segment levels, and allows for an assessment of system connectivity without requiring data like traffic volumes and calculations of the *HCM* Bicycle MMLOS. LTS is well suited for high-level plans such as corridor and transportation system plans, and it is widely used. Notably, the Oregon DOT implemented it in its MMLOS guides (*22*), and a tool to estimate comfort measures to pedestrian and cyclist, called StreetScore + (*23*), was developed using the LTS method.

Deficiency Index. The Deficiency Index (DI) is an alternative to the *HCM* urban streets method (*24*, *25*). The key aspects of this method include using step functions in a modular framework, using deficiency indices, addressing comparability of measures across modes, and incorporating the impact of modes on other modes (*24*). DI classifies features into three groups: physical, operational, and intermodal. Characteristics from the HCM 2010 were included with selected additional components from other research sources to describe each group. Each characteristic is rated by assigning it a deficiency score (DS), which is determined by the user assigning a numerical value based on five ratings: good (0), fair (1), poor (2.5), very poor (4), and extreme (5).

Each characteristic is also assigned a weight (W) value, which was obtained through a survey of transportation professionals working in research, academia, and practice. The weights were rated as follows: very small impact (1), small impact (2), moderate impact (3), large impact (4) and very large impact (5). The DI is obtained by the multiplication of the DS (assigned by the user) and the W values for each characteristic. The DI values are averaged per feature group and then per each mode. The weights of characteristics and DSs can be adjusted by users of the method.

Total Peak-Period Travel Time. The Texas A&M Transportation Institute first introduced the concept of total peak-period travel time (TPPTT) in 2012 for its Urban Mobility Report (*16*). The TPPTT is a multimodal performance measure that represents the door-to-door sum of all travel times during both morning and evening peak periods regardless of mode or travel path (*26*). The method estimates the TPPTT in units of minutes per day. It also determines the travel time index (TTI) obtained as the ratio of the travel time and the free-flow travel time. The key elements included are: number of miles traveled on each roadway classification, free-flow speed, congested speed, and commuter population.

Walk Score®, Bike Score®, and Transit Score®. Walk Score®, Bike Score®, and Transit Score® are web-based tools (www.walkscore.com) developed by Front Seat Management, a software development company based in Seattle, Washington, that focuses on software with civic applications (*27*). The software uses a variety of data sources, including Google, Education.com, Open Street Map, Localeze, city governments, the USGS, public transit agencies, and the U.S. Census. Scores range from 0 to 100. For each address, Walk Score analyzes hundreds of walking routes to nearby amenities, with points awarded based on the distance to amenities in each category. Bike Score is calculated by measuring bicycle infrastructure (e.g., lanes, trails), hills, destinations and road connectivity, and the number of bicycle commuters. Transit Score is estimated through a "usefulness" value calculated by the frequency, type of route (e.g., rail, bus), and distance to the nearest stop on transit routes. The "usefulness" of all nearby routes is summed and normalized to a score between 0 and 100 (*28*).

Case Study

The present research aims to evaluate the multimodal performance of an arterial corridor section in Austin, Texas, using different MMLOS approaches to assessment of pedestrian, bicycle, and transit modes. The methods applied are: *HCM*, *TCQSM*, *USDG*, BEQI/PEQI, LTS, BCI, DI, and Walk Score®, Bike Score®, and Transit Score®. The corridor sector evaluated is a 1.1-mile segment of the total 6.5-mile length of Airport Boulevard, which includes four major signalized intersections: Aldrich Street/Wilshire Boulevard, East 38th ½ Street, Manor Road, and East Martin Luther King Jr. (MLK) Boulevard. Figure 1 presents the corridor along with details of the intersections.

The section from I-35 to Manor Road presents six median-separated lanes designed specifically to accommodate the significant peak-hour traffic near the Airport Boulevard intersection with I-35 (*29*). At Manor Road, the boulevard changes to a four-lane roadway with a continuous center left-turn lane. The posted speed in the area is 45 mph. From I-35 to East 38th ½ Street, the corridor is lower in traffic volume with residential neighborhoods, parks, and landscaped buffers along the roadway, whereas from East 38th ½ Street to MLK Boulevard the corridor is characterized by auto-oriented, service, and warehouse uses.

The analysis is applied at intersection and street-segment levels. Table 1 provides a detailed characterization of the roadway elements of the study. Data sources include a field evaluation, Google Earth, the City of Austin's *Airport Boulevard Corridor Development Program* report (*29*), Capital Metro users' information, and its *System and Service Evaluation* report (*30*). Traffic volumes are based on counts collected on weekdays in

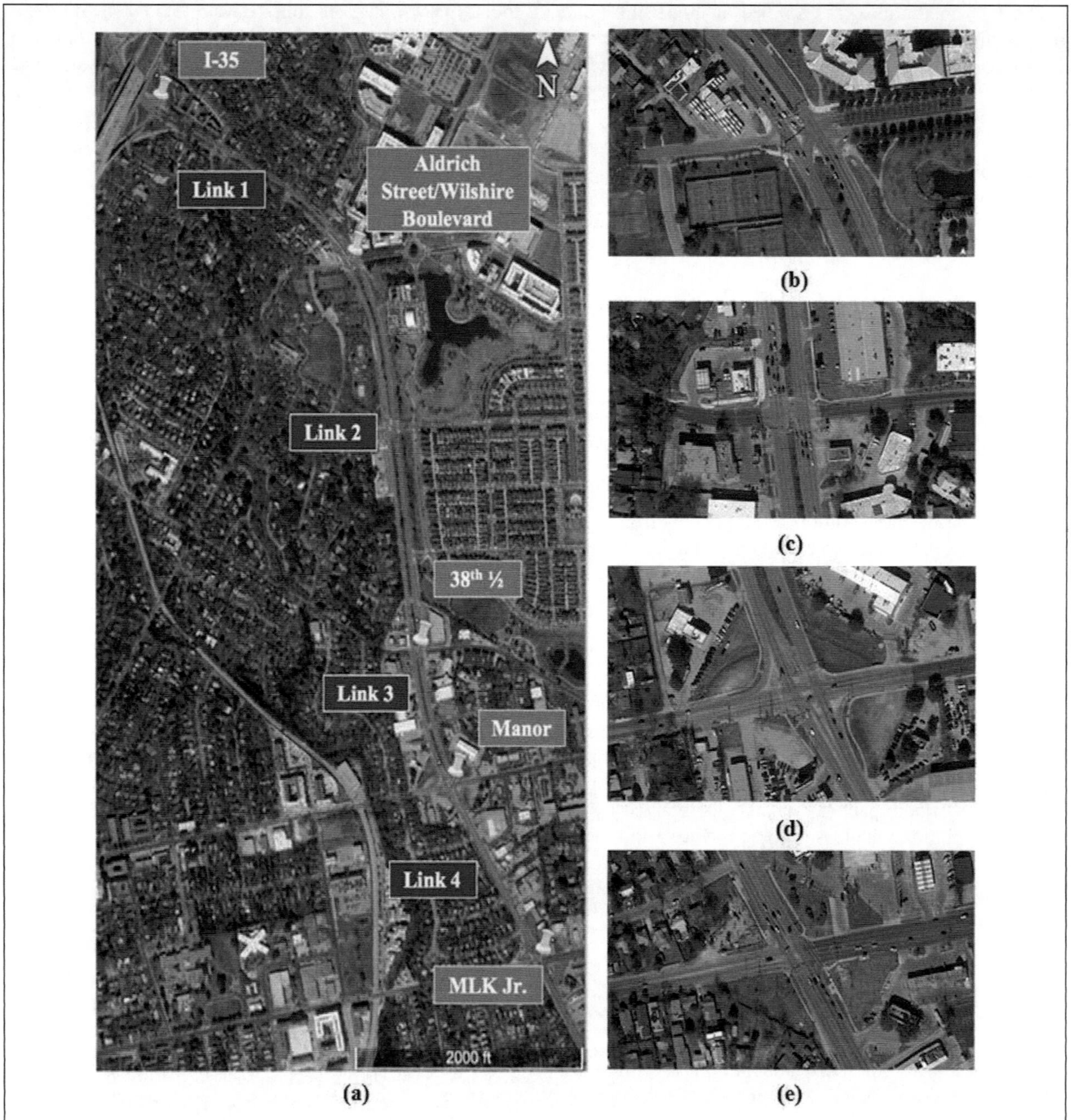

Figure 1. Details of Airport Boulevard and intersections: (*a*) Airport Boulevard corridor; (*b*) Aldrich Street/Wilshire Boulevard; (*c*) East 38th ½ Street; (*d*) Manor Road; (*e*) East MLK Boulevard. (*Source*: Google Earth, 2017).

June 2011 and September 2011 (*29*) that were converted into 2017 volumes using a 3% annual growth rate.

Results and Discussion

The Airport Boulevard evaluation utilized eight MMLOS methodologies. Table 2 provides a comparison of the characteristics evaluated in each. An exception was made for Walk Score®, Bike Score®, and Transit Score®, which were not included because there was no precise information available about the features that they evaluate. In addition, *HCM* and *TCQSM* methodologies complement each other since *HCM* covers pedestrian and bicycle modes and *TCQSM* covers transit. These

Table 1. Airport Boulevard: Description of Segment and Intersection Characteristics

Link	Length (miles)	Sidewalk width (ft) NB	Sidewalk width (ft) SB	Transit routes	Transit stops	Bike lane
Link 1	0.4	4.5	4.5	350, 37, 485, 135	2 stops	None
Link 2	0.6	7.5	7.5	350, 485, 135	No stops	None
Link 3	0.2	6.0	0	350, 485, 135	1 stops	None
Link 4	0.3	6.0	6.0	350, 485, 135	2 stops	None

Intersection	Lanes N	Lanes S	Lanes E	Lanes W	Median-separated approach	Cross-walk treatment	Bike lane
Aldrich Street & Wilshire Boulevard	7	6	2	5	N, S, E	Ladder	E
East 38th ½ Street	7	7	3	3	N, S	Ladder	None
Manor Road	7	6	5	4	N, S	Ladder	None
East MLK Boulevard	6	5	5	5	N, S	Transverse marking	None

Note: NB = Northbound; SB = Southbound; EB = Eastbound; WB = Westbound; N = North approach; S = South approach; E = East approach; W = West approach.

methods have been combined in Table 2, resulting in a total of six columns to address seven MMLOS methods. Table 2 provides information required to evaluate the different modes and presents the differences in data input requirements between the different methods. For example, the table shows how the *HCM* method differs from PEQI/BEQI and DI in that it does not consider features such as cross-walk treatment and Americans with Disabilities Act (ADA) curb ramps. This assessment of the methodologies' differences is an important contribution of the present study.

Application of the methods to the case study is described in the following paragraphs. As per the outlined methodology, the *HCM* intersection LOS approach is applied separately for each intersection leg. The authors have deviated from *HCM* guidelines in calculating an overall intersection score (an average of scores for each intersection leg), a deviation justified so that results can be compared with other methods. Similarly, northbound (NB) and southbound (SB) street segments are evaluated and averaged for a final score. Again, averaging the directional score is justified so that results can be compared with other methodologies. The BEQI and PEQI used are based on the methodology as explained in the users' manuals (*18*, *19*). However, it is important to mention that the BEQI and PEQI have an updated version, Version 2.0. The DI method is used on sections that include both street segments and intersections. The Walk Score®, Bike Score®, and Transit Score® are obtained using the free online resource at www.walkscore.com. However, there is not precise information available about the features that this methodology evaluates to obtain results.

Results are shown in Table 3. The rows describe each methodology and mode, while the columns include the unit analysis (i.e., links and intersections along the corridor). The final results obtained for each mode are comparatively illustrated in Figure 2, where scores from the methodologies are graphically compared to provide a clearer analysis of the results.

Mode Evaluation

This section includes a discussion of the results obtained for each mode.

Pedestrian LOS. The pedestrian LOS results are shown in Figure 2*a*. The methodologies present similar evaluation results along the Airport Boulevard corridor. The Charlotte *USDG* shows the most favorable grades, but it only includes intersections. The intersection evaluation suggests similar pedestrian performance along the corridor. However, this methodology presents a drastic variation from one intersection to other, with Manor Road having the lowest performance. The link evaluation is consistent across the methodologies in showing Link 3 with the lowest grade due to the lack of a pedestrian sidewalk in the NB direction. However, Walk Score® does not capture this link deficiency. Link 2 presents the best conditions for pedestrians due to the recreational area located along the NB direction which allows for a wider sidewalk and buffer. However, only *HCM* and DI show the higher results for this link.

Bicycle LOS. The bicycle LOS results are presented in Figure 2*b*. In this case, the results are more dispersed along the score ranges and present higher variability along the corridor, compared with pedestrian LOS. For instance, Bike Score® is the most favorable, with an

Table 2. Comparison of Characteristics Across Methods Applied

Mode	Characteristic	*HCM/ TCQSM*	*USDG*	BEQI/ PEQI	LTS	BCI	DI
Pedestrian	Presence of sidewalk	X	X				X
	Sidewalk width	X		X			X
	Sidewalk quality						X
	Side street geometry	X	X	X			
	Vehicle volume and speed	X		X			
	Vehicle right turns and permitted lefts	X	X	X			
	Pedestrian volume	X					X
	Pedestrian signal type	X	X	X			
	Presence of physical barrier and buffers	X		X			
	Distance from vehicles	X					X
	Intersection corner radius		X				
	Cross-walk treatment		X	X			X
	Traffic calming feature			X			
	ADA curb ramps			X			X
	Lighting levels			X			
	Visual interest and amenities			X			
	Mid-block and intersection crossing delay	X					X
	Auto, transit, and bicycle impact						X
Bicycle	Auto volumes	X		X		X	
	Auto speeds	X	X	X	X	X	
	Percentage of heavy vehicles	X		X		X	
	Percentage of on-street parking	X		X	X	X	X
	Pavement rating	X					X
	Presence of bike lane or paved shoulder	X	X	X	X	X	X
	Width of bicycle lane	X		X	X		X
	Width of outside lane	X	X		X		X
	Bike lane blockage				X		
	Presence of physical barrier and buffers				X		
	Intersection crossing distance	X	X				
	Right turns on red	X	X	X		X	
	Right-turn lane length				X		
	Traffic calming features			X			
	Bicycle parking			X			
	Connection to on-street lanes		X	X			
	Line of sight, street slope, lighting			X			
	Residential development					X	
	Auto, transit, and pedestrian impact						X
Transit	Frequency of service	X					X
	Average transit travel speed	X					X
	Average excess wait time	X					
	Bus stop amenities	X					X
	Bus load factor	X					X
	Span of service						X
	Auto volumes and speed	X					
	Sidewalk width and connection to stop	X					
	Outside lane, shoulder, bike lane width	X					
	Number of travel lanes	X					
	Pedestrian crossing difficulty						X
	Accessibility by bicycle						X
	Delay caused by auto						X

average score of 93. Other methods provide lower scores, mainly due to the lack of a bicycle lane. LTS presents the lowest score of the range across all units of study (street segments and intersections). BCI and *USDG* provide similar results with low variability along the corridor, although one method evaluates only street segments (BCI) and the other only intersections (*USDG*). However, both methods use similar inputs, e.g., auto speeds, percentage of on-street parking, and the presence of bike lanes (refer to Table 2).

The *HCM* and BEQI scores present high variation between intersections and links. The street segment

Table 3. MMLOS Methodologies Results

Method	Mode	Link 1	Airport Blvd & Wilshire/ Aldrich	Link 2	Airport Blvd & E 38th 1/2 St	Link 3	Airport Blvd & Manor	Link 4	Airport Blvd & E MLK Jr Blvd	Average
HCM	Pedestrian	D (3.66)	C (2.99)	B (2.53)	C (2.86)	E (4.41)	C (3.11)	C (3.27)	C (3.12)	C (3.24)
	Bicycle	E (4.73)	C (3.32)	E (4.61)	B (2.35)	E (4.59)	C (2.98)	E (4.82)	C (2.99)	D (3.80)
TCQSM	Transit	B (2.50)	–	B (2.21)	–	B (2.76)	–	B (2.73)	–	B (2.55)
Charlotte *USDG*	Pedestrian	–	B (85)	–	B (78)	–	C (63)	–	C (69)	B$^-$(74)
	Bicycle	–	E$^+$ (35)	–	E$^+$ (34)	–	E (28)	–	E (25)	E (31)
BCI	Bicycle	F (5.83)	–	D (4.20)	–	D (4.39)	–	D (4.33)	–	E (4.69)
PEQI	Pedestrian	62	71	70	67	48	69	57	61	63
BEQI	Bicycle	40	11	62	0	54	0	52	0	27
LTS	Bicycle	LTS4	LTS4	LTS4	LTS4	LTS4	LTS4	LTS4	LTS4	LTS4
DI	Pedestrian	5.93		1.54		7.25		4.44		4.79
	Bicycle	9.16		7.54		8.02		7.91		8.16
	Transit	3.29		4.48		4.13		3.07		3.74
Walk Score®	Pedestrian	60	60	65	64	67	67	59	53	62
Bike Score®	Bicycle	94	93	92	94	97	97	92	83	93
Transit Score®	Transit	48	46	46	46	46	46	48	48	47

evaluation of BEQI includes approximately 20 characteristics. However, the method only considers three aspects for assessment of intersections: left-turn bicycle lane, dashed-intersection bicycle lane, and no-turn-on-red signs. For this reason, the score value is low if the intersection does not have any of them. The *HCM* street segment grades seem to be highly influenced by the lack of a bicycle lane, high auto speeds, and the presence of high traffic volumes. These limitations do not drastically affect the LOS result at intersection level. Also, the *HCM* intersection scores include the average of the four legs, and the crossing streets present significantly lower traffic volumes and better conditions for cyclists than Airport Boulevard.

Transit LOS. Transit LOS results are shown in Figure 2*c*. *TCQSM* and the DI present similar results. The main difference is observed in Link 2, which is a 0.6-mile segment that does not have transit stops. The positive higher value from the *TCQSM* can be related to the high pedestrian performance conditions because the method includes the interaction with this mode in the results. Transit Score®, which gives more weight to the "usefulness" of the routes based on frequency, type of route, and distance to the nearest stop on the route, presents lower performance scores.

Comparison of Methodologies

The methodologies are analyzed based on their applicability to the arterial corridor evaluation. The Charlotte *USDG* provides a detailed evaluation of intersections, and its procedure can be easily applied by the user with a spreadsheet tool. However, the *USDG* method does not evaluate street segments, which tend to be critical for pedestrians and cyclists in a corridor evaluation, as noted in the other methods' results. Similarly, the BCI provides a useful tool for street segments but does not evaluate intersections. Intersections are critical, especially in terms of pedestrian and cyclist safety.

PEQI and BEQI offer a checklist for field evaluation that can be applied by a user with minimal training. It provides results comparable with the other methods. However, the main limitation of this method is its bicycle intersection score because it only considers three aspects, thus the results are not robust. The LTS method is considered as an evaluation resource that does not require intense data collection. Although the method provides only four ranges based on compatibility of the streets to different types of cyclists, it is considered a practical tool that gives easy-to-interpret results. However, in a multimodal evaluation, it would be difficult to compare bicycle results with other modes performance.

HCM provides a pedestrian, bicycle, and transit (*TCQSM*) evaluation that considers the interaction of modes. This approach can evaluate intersections and street segments separately, and provides results comparable across modes. The DI method provides an accurate tool to compare across modes and sections. However, this approach is subjective since the scale to evaluate the characteristics does not provide a standard classification.

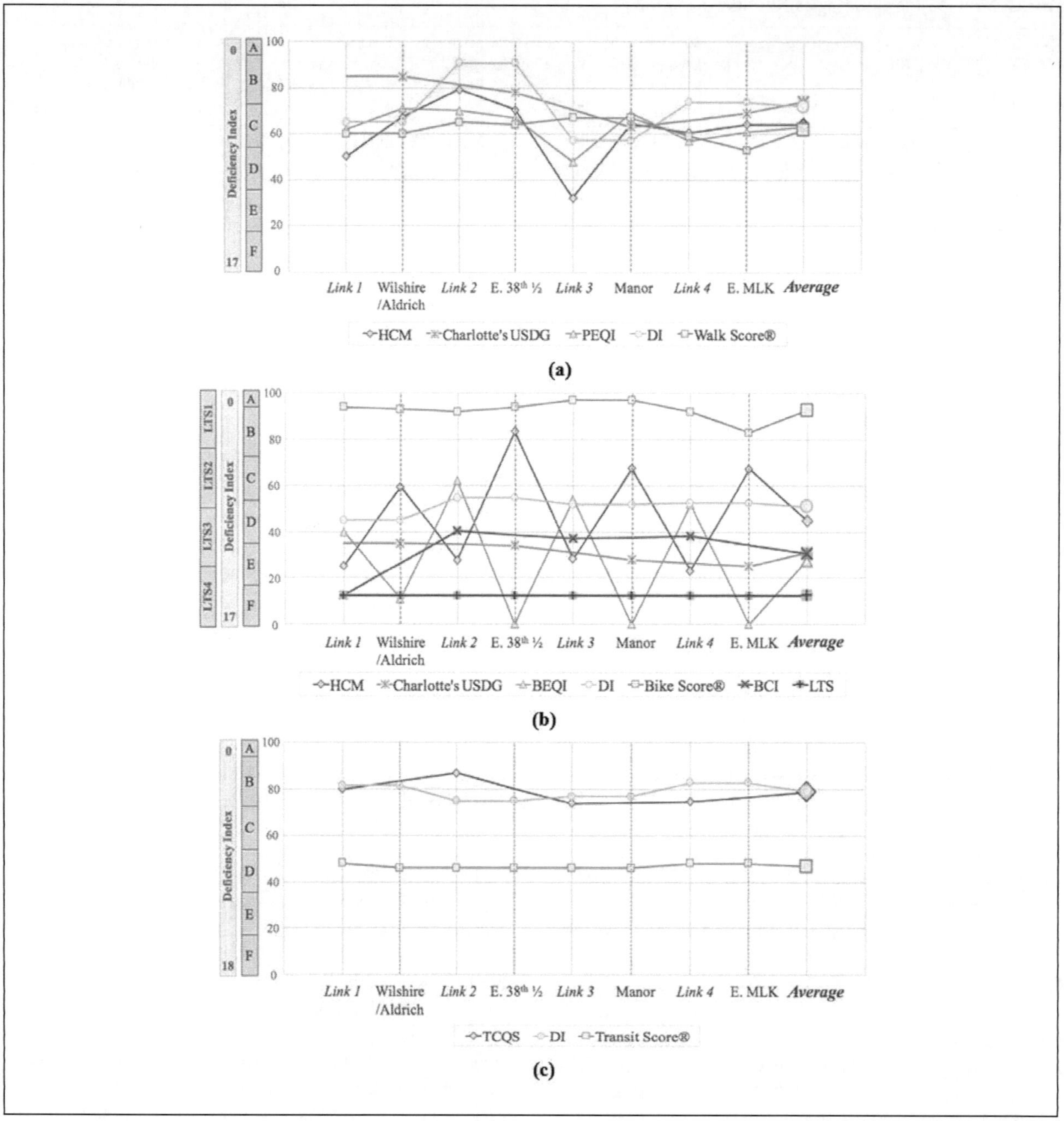

Figure 2. Comparison of MMLOS methodologies: (*a*) Pedestrian Level of Service; (*b*) Bicycle Level of Service; (c) Transit Level of Service.

For instance, when valuing the width of sidewalks on a scale from "good" to "extreme" there are no corresponding width measurements, and evaluation requires that the user has the technical experience and judgement to make such an assessment. Thus, DI is recommended when applied with other methods that can provide a more standard evaluation.

The Walk Score®, Bike Score®, and Transit Score® method presents an assessment of the accessibility of the location to pedestrians, cyclists, and transit users. However, the method presents higher scores on a section where infrastructure is inadequate, e.g., lack of sidewalk and bicycle lane. The method seems to average the characteristics of the surrounding areas, which in this case are more favorable than the corridor. Additionally, the methodology is not explained broadly, and the method cannot be applied directly by the user. A score is given only based on the location. Thus, this methodology is

Table 4. Applicability of MMLOS Methodologies to Corridor Analysis

Method	Mode	Pros	Cons
HCM	Pedestrian and bicycle	• Evaluates both intersection and links • Considers interaction of modes • Strong research background	• Not easy to apply • Requires training and technical knowledge • Requires detailed data collection
TCQSM	Transit	• Easy to apply using the spreadsheet tool • Considers interaction of modes • Strong research background	• Requires detailed data collection
Charlotte USDG	Pedestrian and bicycle	• Easy to apply using the spreadsheet tool • Detailed intersection assessment	• Does not evaluate link segments
PEQI/BEQI	Pedestrian and bicycle	• Evaluates both intersection and links • Easy to apply • Requires minimal basic training	• Bicycle intersection assessment only considers three features
LTS	Bicycle	• Easy to apply • Evaluates both intersection and links • Does not requires intense data collection	• Does not evaluate pedestrian and transit
BCI	Bicycle	• Easy to apply • Does not require intense data collection	• Does not evaluate intersections • Does not evaluate pedestrian and transit
DI	Pedestrian, bicycle, and transit	• Evaluates pedestrian, bicycle, and transit using comparable measures • Considers interaction of modes • Can be used in conjunction with other methods	• Requires technical knowledge • Subjective scale of application (not and standard evaluation)
Walk Score® Bike Score® Transit Score®	Pedestrian, bicycle, and transit	• Easy to apply • Evaluates pedestrian, bicycle, and transit using comparable measures • Does not require data collection process • Evaluates both intersection and links	• Not sensitive to infrastructure deficiencies (e.g. lack of bike lane or sidewalk) • Methodology not reproducible

not recommended for the evaluation of roadway infrastructure assessment along a corridor.

Table 4 summarizes the pros and cons of applying each methodology to a corridor assessment.

Corridor Multimodal Performance

The main objective of this study was to evaluate the multimodal performance of an arterial corridor using current methodologies. Results showed variability in LOS between street segments and intersections, mainly for the bicycle mode. This is important because an average of links and intersections across the corridor (as presented in Figure 2) shows a rough estimate of LOS that does not include this variability. Thus, it may be relevant to consider separation of results in terms of intersections or street segments.

It is also apparent that an overall multimodal performance number should not be used. Combining LOS across modes would disguise the disparities in the perceptions of the quality of service of the independent mode. The main reasons not to combine modes are: (a) there is a not professionally and scientifically proven accepted method; (b) the weight of each mode should be established based on specific use, which will depend on relative importance, policy goals, or other criteria; (c) the purpose and travel patterns are different per mode.

Summary and Conclusions

This research study explored the current MMLOS methodologies with the objective of evaluating the multimodal performance of arterial corridors. A case study of a corridor section in Austin, Texas, is analyzed using different approaches: *HCM*, *TCQSM*, Charlotte *USDG*, BEQI/

PEQI, LTS, BCI, DI, and Walk Score®, Bike Score®, and Transit Score®. The evaluated section comprised 1.5 miles of the entire 6.5-mile corridor, with the analysis applied at intersection and street-segment levels using four links and four signalized intersections that include pedestrian, bicycle, and transit modes.

A summary of this study is provided below:

- Through the literature about the state of the practice, it was possible to describe six MMLOS procedures that use letter-metric grades and five approaches that use different score measures. A review of the most relevant methods for an understanding of the use and limitations of each method was presented.
- The case study provided a comparison of methodologies that allows for the evaluation of the approaches according to their applicability to multimodal arterial corridors. The methods must analyze both street segments and intersections. It is necessary to provide comparability across modes. Thus, the recommended methodologies are *HCM*, *TCQSM*, and DI. However, it is recommended to combine the DI approach with other methodologies to provide a more standard evaluation.
- The study provides information that can help practitioners select an appropriate methodology for evaluating mode performance based on available information. Table 2 presents a comparison of the information necessary to apply the different methodologies and provides an overall understanding of the features included in each. Table 4 summarizes the pros and cons of applying each method to a corridor assessment.

The major findings and conclusions from this study are:

- *HCM* provides a MMLOS assessment that is suitable for a corridor evaluation. However, its application requires user training and significant data collection.
- *TCQSM* complements *HCM* and provides assessment of the transit mode for corridor evaluation.
- Charlotte *USDG* is not recommended for corridor evaluation because it does not assess link segments. However, it does provide a detailed evaluation of intersections.
- PEQI and BEQI offer an appropriate corridor assessment. However, the analysis of bicycle LOS at intersections does not provide a sufficiently robust evaluation and may cause significantly lower values than are found using other methodologies.
- LTS and BCI evaluate bicycle LOS with easy application and minimal data requirements. Application to a MMLOS corridor assessment requires the use of complementary pedestrian and transit methods.
- DI is the most robust method. It can provide comparison across modes and consider interaction between modes. However, it requires technical knowledge and its application is subjective to user expertise. It is recommended to be used with other methods.
- Walk Score®, Bike Score®, and Transit Score® provide comparability across modes and analyze both intersections and street segments. However, the method seems to be insensitive to important roadway infrastructural features in a multimodal assessment (e.g., lack of sidewalks and bike lanes). Additionally, the methodology is not widely explained, and it is not possible to reproduce the score without using the proprietary web tool. Thus, this method is not recommended for arterial corridor evaluation.
- An overall MMLOS that includes multiple modes (i.e., automobile, transit, pedestrian, and bicycle) is not provided by any of the methods. The multimodal analysis should be applied separately for each mode. A further aggregation of the results to provide one overall score requires researcher judgement in terms of weights for the modes, and therefore biases results by assigning relative importance to individual modes through the act of aggregation.

A complementary effort to this paper was the development of an MMLOS spreadsheet tool for the City of Austin to use in assessing LOS for transportation corridors. For that tool, the authors recommended applying *HCM* and *TCQSM* methodologies to assess the corridors. The methodologies were applied as directed, with minor deviations to tailor the analysis to the corridor in question, as well as to compare analysis results across scenarios more easily.

Author Contributions

The authors confirm contribution to the paper as follows: study conception and design: Natalia Zuniga-Garcia, Heidi W. Ross, and Randy B. Machemehl; data collection: Natalia Zuniga-Garcia; analysis and interpretation of results: Natalia Zuniga-Garcia, Heidi W. Ross, and Randy B. Machemehl; draft manuscript preparation: Natalia Zuniga-Garcia. All authors reviewed the results and approved the final version of the manuscript.

References

1. Michael, A. C., and E. C. Yamamoto. Level of Service Concepts in Multimodal Environments. In *Traffic Engineering Handbook: Institute of Transportation Engineers* (A. Pande, and B. Wolshon, eds.), 7th ed., John Wiley & Sons, Hoboken, NJ, 2015, pp. 149–176.
2. Dowling, R., A. Flannery, B. Landis, T. Petritsch, N. Rouphail, and P. Ryus. Multimodal Level of Service for Urban Streets. *Transportation Research Record: Journal of the Transportation Research Board*, 2008. 2071: 1–7.
3. *Highway Capacity Manual 2010*. Transportation Research Board of the National Academies, Washington, D.C., 2010.
4. *Highway Capacity Manual: A Guide for Multimodal Mobility Analysis*, 6th ed. Transportation Research Board, Washington, D.C., 2016.
5. Brozen, M., H. Huff, R. Ligget, and M. Smart. Exploration and Implications of Multimodal Street Performance Metric: What's a Passing Grade? Report No. UCTC-FR-2014-09. University of California Transportation Center, Los Angeles, CA, 2014. http://www.lewis.ucla.edu/wp-content/uploads/sites/2/2014/09/EXPLORATION-AND-IMPLICATIONS-OF-MULTIMODAL-STREET-PERFORMANCE-METRICS.pdf. Accessed November 14, 2017.
6. Landis, B., V. Vattikuti, and M. Brannick. Real-Time Human Perceptions Toward a Bicycle Level of Service. *Transportation Research Record: Journal of the Transportation Research Board*, 1997. 1578: 119–125.
7. Landis, B., V. Vattikuti, R. Ottenberg, D. McLeod, and M. Guttenplan. Modeling the Roadside Walking Environment: Pedestrian Level of Service. *Transportation Research Record: Journal of the Transportation Research Board*, 2001. 1773: 82–88.
8. Landis, B., V. Vattikuti, R. Ottenberg, T. Petritsch, M. Guttenplan, and L. Crider. Intersection Level of Service for the Bicycle Through Movement. *Transportation Research Record: Journal of the Transportation Research Board*, 2003. 1828: 101–106.
9. Kittelson & Assoc., Inc., Parsons Brinckerhoff, Inc., KFH Group, Inc., Texas A&M Transportation Institute, and Arup. *Transit Cooperative Highway Research Program*, 3rd ed. Report No. 165. Transit Capacity and Quality of Service Manual, Transportation Research Board, Washington, D.C., 2013.
10. *Level of Service Case Studies Evolving Use of Level of Service Metrics in Transportation Analysis*. FHWA, U.S. Department of Transportation, Washington D.C., 2017. https://www.transportation.gov/office-policy/transportation-policy/level-service-case-studies. Accessed November 14, 2017.
11. *2013 Quality/Level of Service Handbook*. State of Florida Department of Transportation, Tallahassee, Florida, 2013. http://www.fdot.gov/planning/systems/programs/SM/los/pdfs/2013%20QLOS%20Handbook.pdf. Accessed November 14, 2017.
12. Dowling, R., P. Ryus, B. Schroeder, M. Kyte, F. T. Creasey, N. Rouphail, A. Hajbabaie, and D. Rhoades. *NCHRP Report 825: Planning and Preliminary Engineering Applications Guide to the Highway Capacity Manual*. Transportation Research Board of the National Academies, Washington, D.C., 2016.
13. *Urban Street Design Guidelines*. City of Charlotte, Charlotte, NC, 2007. http://charlottenc.gov/Transportation/PlansProjects/Documents/USDG%20Full%20Document.pdf. Accessed November 14, 2017.
14. *2009 National Award for Smart Growth Achievement*. U.S. Environmental Protection Agency, Washington D.C., 2009. https://www.epa.gov/smartgrowth/2009-national-award-smart-growth-achievement-booklet. Accessed November 14, 2017.
15. *Multimodal Transportation Level of Service Manual*. Fort Collins, CO, 1997. https://www.larimer.org/sites/default/files/apdxh.pdf. Accessed November 14, 2017.
16. Lasley, P. Developing Multimodal Performance Measures: Understanding What They Are and How to Develop Them. Presented at 95th Annual Meeting of the Transportation Research Board, Washington, D.C., 2016.
17. Harkey, D. L., D. W. Reinfurt, M. Knuiman, and A. Sorton. *Development of the Bicycle Compatibility Index: A Level of Service Concept*. FHWA-RD-98-072. FHWA, U.S. Department of Transportation, Washington, D.C., 1998. https://safety.fhwa.dot.gov/tools/docs/bcifinalrpt.pdf. Accessed November 14, 2017.
18. *The Pedestrian Environmental Quality Index (PEQI): An Assessment of the Physical Condition of Streets and Intersections*. San Francisco Department of Public Health, San Francisco, CA, 2008.
19. *Bicycle Environmental Quality Index Report (BEQI)*. San Francisco Department of Public Health, San Francisco, CA, 2009.
20. Necessary, M. *Developing an Infrastructure Informed Index for Pedestrians and Bicyclists*. Technical Report No. D-STOP/2016/116. University of Texas at Austin, Austin, TX, 2016. http://ctr.utexas.edu/wp-content/uploads/DSTOP116.pdf. Accessed November 14, 2017.
21. Mekuria, M. C., P. G. Furth, and H. Nixon. *Low-Stress Bicycling and Network Connectivity*. Report 11-19. Mineta Transportation Institute, San Jose, CA, 2012.
22. *Analysis Procedures Manual: Multimodal Analysis (Chap. 14)*. State of Oregon Department of Transportation, Salem, OR, 2016, pp. 14-1–14-73. http://www.oregon.gov/ODOT/TD/TP/APM/APMv2_Ch14.pdf. Accessed November 14, 2017.
23. Fehr & Peers Transportations Consultants. *Multi-Modal Level of Service Toolkit: Bicycling Level of Traffic Stress*. Washington D.C., 2014. http://asap.fehrandpeers.com/wp-content/uploads/2014/08/MMLOS-Tool-Level-of-Traffic-Stress.pdf. Accessed November 14, 2017.
24. Kastenhofer, I. O. *Multimodal Assessment of Recurrent and Non-recurrent Conditions on Urban Streets*. PhD dissertation, Virginia Polytechnic Institute and State University, 2014.
25. Kastenhofer, I. O., A. G. Hobeika, and M. M. Abbas. A Method to Analyze Multimodal Conditions on Congested Urban and Suburban Arterials. *ITE Journal*, Vol. 87, No. 4, 2017, pp. 42–48.
26. Lasley, P., T. Lomax, W. Eisele, and D. Schrank. Developing a Total Peak Period Travel Time Performance Measure: An Updated Concept Paper. *Transportation Research*

Record: Journal of the Transportation Research Board, 2014. 2420: 15–22.
27. Duncan, D. T., J. Aldstadt, J. Whalen, S. J. Melly, and S. L. Gortmaker. Validation of Walk Score® for Estimating Neighborhood Walkability: An Analysis of Four US Metropolitan Areas. *International Journal of Environmental Research and Public Health*, Vol. 8, No. 11, 2011, pp. 4160–4179.
28. *Walk Score Methodology*. Front Seat, Seattle, WA, 2017. http://www.walkscore.com/methodology.shtm. Accessed November 14, 2017.
29. *Report: Airport Boulevard Corridor Development Program*. City of Austin Transportation Department, Austin, TX, 2014. http://austintexas.gov/department/airport-boulevard-mobility-program. Accessed November 14, 2017.
30. Transportation Management & Design, Incorporated. *Connections 2025 A Transit Plan for the Future Report: System and Service Evaluation*. Transportation Management & Design, Incorporated, Austin, TX, 2016. http://connections2025.org/wp-content/uploads/2016/07/Capital-Metro-Service-Review.pdf. Accessed November 14, 2017.

The Standing Committee on Highway Capacity and Quality of Service (AHB40) peer-reviewed this paper (18-00844).

Sage Publications USA
sagepub.com

ISBN 978-1-5108-7733-7

CURRAN ASSOCIATES INC.
proceedings
.com

2nd International Conference on Mechatronics and Electrical Systems (ICMES 2017)

IOP Conference Series: Materials Science and Engineering
Volume 339

Wuhan, China
15 – 17 December 2017

IOP
Institute of Physics